Mapping and Monitoring Forest Cover

Mapping and Monitoring Forest Cover

Editor

Russell G. Congalton

MDPI • Basel • Beijing • Wuhan • Barcelona • Belgrade • Manchester • Tokyo • Cluj • Tianjin

Editor
Russell G. Congalton
University of New Hampshire
USA

Editorial Office
MDPI
St. Alban-Anlage 66
4052 Basel, Switzerland

This is a reprint of articles from the Special Issue published online in the open access journal *Forests* (ISSN 1999-4907) (available at: https://www.mdpi.com/journal/forests/special_issues/mapping_cover).

For citation purposes, cite each article independently as indicated on the article page online and as indicated below:

LastName, A.A.; LastName, B.B.; LastName, C.C. Article Title. *Journal Name* **Year**, *Volume Number*, Page Range.

ISBN 978-3-0365-2043-8 (Hbk)
ISBN 978-3-0365-2044-5 (PDF)

Cover image courtesy of Russ Congalton

Contents

About the Editor

Russell G. Congalton is Professor of Remote Sensing and GIS in the Department of Natural Resources and the Environment at the University of New Hampshire. He is responsible for teaching courses in geospatial analysis including The Science of Where, Remote Sensing, Photogrammetry and Image Interpretation, Digital Image Processing, and Geographic Information Systems. Russ has authored or coauthored more than 200 papers and conference proceedings. His papers have won awards seven times including: 1994 ASPRS John I. Davidson Award for Practical Papers (2nd Prize), 1996 ESRI Award for Best Scientific Paper in Geographic Information Systems (3rd Prize), 1998 ASPRS John I. Davidson Award for Practical Papers (1st Prize), 1998 ESRI Award for Best Scientific Paper in Geographic Information Systems (2nd Prize), 2014 ERDAS Award for Best Scientific Paper in Remote Sensing (3rd Prize), and 2016 ERDAS Award for Best Scientific Paper in Remote Sensing (1st Prize). He is the author of thirteen book chapters, co-editor of a book on spatial uncertainty in natural resource databases titled Quantifying Spatial Uncertainty in Natural Resources: Theory and Applications for GIS and Remote Sensing, and is co-author of five books including Assessing the Accuracy of Remotely Sensed Data: Principles and Practices (1st, 2nd, and 3rd editions published by CRC Press) and Imagery and GIS: Best Practices for Extracting Information from Imagery, published by ESRI in 2017. Russ served as President of the American Society for Photogrammetry and Remote Sensing (ASPRS) in 2004–2005, as the National Workshop Director for ASPRS from 1997 to 2008, and as Editor-in-Chief of Photogrammetric Engineering and Remote Sensing from 2008 to 2016. He was elected a Fellow in 2007 and an Honorary Member in 2016. In 2005, Dr. Congalton was awarded the inaugural Graduate Faculty Mentor Award from the University of New Hampshire for his dedication to graduate education and his graduate students. He has served as primary mentor and advisor for over 60 graduate students during his career.

Dr. Congalton received a B.S. (Natural Resource Management) from Rutgers University in 1979. He earned an M.S. (1981) and a Ph.D. (1984) in remote sensing and forest biometrics from Virginia Tech. In addition to his academic position, Russ served as Chief Scientist of Pacific Meridian Resources from its founding in 1988 until 2000 and then as Chief Scientist of Space Imaging Solutions from 2000 to 2004. From 2004 to 2015, Russ held the position of Senior Technical Advisor with the Solutions Group of Sanborn, the oldest mapping company in the US. Dr. Congalton was the Remote Sensing/Land Cover Principal Investigator of the NSF GLOBE Program, a scientist–teacher–student environmental education and research partnership involving over 100 countries and 15,000 schools from 1995 to 2007. Much of the work in this project focused on developing scientific protocols and learning activities for student understanding of land cover mapping and remote sensing. Currently, he is the director of New Hampshire View—a part of the AmericaView Consortium dedicated to promoting the use of remote sensing and other geospatial technologies throughout the US and has served as member, secretary, vice-chair, and chair of the AmericaView Board of Directors.

Editorial

Mapping and Monitoring Forest Cover

Russell G. Congalton

Department of Natural Resources & the Environment, University of New Hampshire, 56 College Road, 114 James Hall, Durham, NH 03824, USA; Russ.Congalton@unh.edu

Citation: Congalton, R G. Mapping and Monitoring Forest Cover. *Forests* **2021**, *12*, 1184. https://doi.org/10.3390/f12091184

Received: 24 August 2021
Accepted: 27 August 2021
Published: 1 September 2021

Publisher's Note: MDPI stays neutral with regard to jurisdictional claims in published maps and institutional affiliations.

Our Earth consists of approximately 70 percent water and 30 percent land. Of the land, approximately 31 percent is forested. Forests provide incredible benefits to all the living creatures on Earth. They provide a diverse ecosystem that is home to countless species of plants and animals. Forests are incredibly diverse—from the boreal forests in the north to the tropical forests of the Amazon, from the pine plantations of New Zealand to the mixed deciduous forests of the northeastern United States and the coniferous forests of Europe, and everywhere in between. Forests are used by humans as a source of heat, building materials, paper, and food. Forests also provide an invaluable place for recreation. Finally, they provide a place for storing carbon and for releasing oxygen. Clearly, forests must be mapped and monitored so that we can effectively manage our forests for a sustainable future.

The use of remotely sensed imagery and other geospatial technologies holds the key to our effective mapping and monitoring of our forests. While collecting samples of ground data is still important for the development and validation of this mapping and monitoring, remote sensing provides a total enumeration of the entire Earth on a repeated basis. Remotely sensed imagery can be collected at a variety of spatial, spectral, and temporal resolutions. Satellites in orbit above the Earth collect imagery continuously and revisit the same spot at regular intervals (every 16 days, every 3 days, twice a day). These satellites tend to have sensors that range from just the visible portions of the electromagnetic spectrum into the near and middle infrared, and some even collect thermal data. The spatial resolution tends to be coarse to moderate (1 km to 30 m), but newer sensors have achieved higher spatial resolutions (10 m to 50 cm). Aircraft can fly over an area repeatedly in a day or over any desired period and can include very high spatial resolution digital cameras or other sensors. Most recently, unmanned aerial systems (UAS) can be used to acquire imagery repeatedly at very high spatial resolutions (a few cms) by flying very low to the ground. Advances in computer processing and algorithms for converting the raw remotely sensed imagery into forest maps or other land cover maps have improved greatly over the last 30 years and continue to do so. The combination of the right imagery with the best processing allows remote sensing scientists to map and monitor our forests far better than we have ever been able to in the past.

This Special Issue comprises six papers that clearly demonstrate the power of using remotely sensed imagery to map and monitor our forests. These papers represent a wide range of examples and emphasize the importance of spatial, spectral, and temporal resolutions provided by a variety of remotely sensed imagery. The paper by Guo et al. [1] demonstrates the benefits of using moderate-resolution imagery to monitor forest change over time and for a large area. The papers by Aljahdali et al. [2] and Jallat et al. [3] emphasize specific critical forest types (mangrove and juniper forests) that are especially important to map and monitor to maintain these niche ecosystems. The paper regarding large juniper forests also introduces the concept of carbon sequestration, which is particularly important to our environment today [3]. The final three papers deal with higher spatial resolution remotely sensed imagery [4–6]. The paper by Ganz et al. [4] demonstrates the benefits of having a detailed forest map of a German forest created from high spatial resolution imagery. They then introduce the concept of comparing or relating what can be determined on the detailed map created from the remotely sensed imagery with their forest inventory data collected on the ground. The paper by Lister et al. [5] continues this important theme. This review paper provides a discussion of the efficiencies that the United States Forest Service employs by combining the National Forest Inventory and Analysis (FIA) data and remote sensing [5]. Many experiences and suggestions provided by this paper can help other countries to more effectively and efficiently map and monitor their forests. The final paper by Gu et al. [6] brings the analysis of forests down to the individual tree level. This paper reports on efforts using a UAS to collect imagery that can be used to delineate individual tree crowns and emphasizes some of the computer processing and algorithm developments that have made such techniques possible [6].

Funding: This research received no external funding.

Conflicts of Interest: The author declares no conflict of interest.

References

1. Guo, Y.; Zeng, J.; Wu, W.; Hu, S.; Liu, G.; Wu, L.; Bryant, C. Spatial and temporal changes in vegetation in the Ruoergai Region, China. *Forests* **2021**, *12*, 76. [CrossRef]
2. Aljahdali, M.; Munawar, S.; Khan, W. Monitoring mangrove forest degradation and regeneration: Landsat time series analysis of moisture and vegetation indices at Rabigh Lagoon, Red Sea. *Forests* **2021**, *12*, 52. [CrossRef]
3. Jallat, H.; Khokhar, M.; Kudus, K.; Nazre, M.; Saqib, N.; Tahir, U.; Khan, W. Monitoring carbon stock and land-use change in 5000-year-old juniper forest stand of Ziarat, Balochistan, through a synergistic approach. *Forests* **2021**, *12*, 51. [CrossRef]
4. Ganz, S.; Adler, P.; Kandler, G. Forest cover mapping based on a combination of aerial images and Sentinel-2 satellite data compared to National Forest Inventory data. *Forests* **2020**, *11*, 1322. [CrossRef]
5. Lister, A.; Andersen, H.; Frescino, T.; Gatziolis, D.; Healey, S.; Heath, L.; Liknes, G.; McRoberts, R.; Moisen, G.; Nelson, M.; et al. Use of remote sensing data to improve the efficiency of national forest inventories: A case study from the United States National Forest Inventory. *Forests* **2020**, *11*, 1364. [CrossRef]
6. Gu, J.; Grybas, H.; Congalton, R. A comparison of forest crown delineation from unmanned aerial imagery using canopy height models vs. spectral lightness. *Forests* **2020**, *11*, 605. [CrossRef]

 *forests*

Article

Spatial and Temporal Changes in Vegetation in the Ruoergai Region, China

Yahui Guo [1,2,†], Jing Zeng [3,†], Wenxiang Wu [1,*], Shunqiang Hu [3], Guangxu Liu [4], Linsheng Wu [5] and Christopher Robin Bryant [6,7]

1 Key Laboratory of Land Surface Pattern and Simulation, Institute of Geographic Sciences and Natural Resources Research, Chinese Academy of Sciences, Beijing 100101, China; guoyh@lreis.ac.cn
2 College of Water Sciences, Beijing Normal University, Beijing 100875, China
3 College of Resource Environment and Tourism, Capital Normal University, Beijing 100048, China; 2200902168@cnu.edu.cn (J.Z.); hushunqiang8@gmail.com (S.H.)
4 School of Geography and Environmental Engineering, Gannan Normal University, Ganzhou 341000, China; liren4z@hotmail.com
5 International Institute for Earth System Sciences, Nanjing University, Nanjing 210023, China; wulinsheng@cnu.edu.cn
6 School of Environmental Design and Rural Development, University of Guelph, Guelph, ON N1G2W5, Canada; chris.r.bryant@umontreal.ca
7 Département de Géographie, Université de Montréal, Montréal, QC H2V2B8, Canada
* Correspondence: wuwx@igsnrr.ac.cn
† These authors contributed equally to this paper.

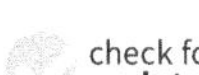

check for
updates

Citation: Guo, Y.; Zeng, J.; Wu, W.; Hu, S.; Liu, G.; Wu, L.; Bryant, C.R. Spatial and Temporal Changes in Vegetation in the Ruoergai Region, China. *Forests* **2021**, *12*, 76. https://doi.org/10.3390/f12010076

Received: 15 December 2020
Accepted: 8 January 2021
Published: 11 January 2021

Publisher's Note: MDPI stays neutral with regard to jurisdictional claims in published maps and institutional affiliations.

Abstract: Timely monitoring of the changes in coverage and growth conditions of vegetation (forest, grass) is very important for preserving the regional and global ecological environment. Vegetation information is mainly reflected by its spectral characteristics, namely, differences and changes in green plant leaves and vegetation canopies in remote sensing domains. The normalized difference vegetation index (NDVI) is commonly used to describe the dynamic changes in vegetation, but the NDVI sequence is not long enough to support the exploration of dynamic changes due to many reasons, such as changes in remote sensing sensors. Thus, the NDVI from different sensors should be scientifically combined using logical methods. In this study, the Global Inventory Modeling and Mapping Studies (GIMMS) NDVI from the Advanced Very High Resolution Radiometer (AVHRR) and Moderate-resolution Imaging Spectroradiometer (MODIS) NDVI are combined using the Savitzky–Golay (SG) method and then utilized to investigate the temporal and spatial changes in the vegetation of the Ruoergai wetland area (RWA). The dynamic spatial and temporal changes and trends of the NDVI sequence in the RWA are analyzed to evaluate and monitor the growth conditions of vegetation in this region. In regard to annual changes, the average annual NDVI shows an overall increasing trend in this region during the past three decades, with a linear trend coefficient of 0.013/10a, indicating that the vegetation coverage has been continuously improving. In regard to seasonal changes, the linear trend coefficients of NDVI are 0.020, 0.021, 0.004, and 0.004/10a for spring, summer, autumn, and winter, respectively. The linear regression coefficient between the gross domestic product (GDP) and NDVI is also calculated, and the coefficients are 0.0024, 0.0015, and 0.0020, with coefficients of determination (R^2) of 0.453, 0.463, and 0.444 for Aba, Ruoergai, and Hongyuan, respectively. Thus, the positive correlation coefficients between the GDP and the growth of NDVI may indicate that increased societal development promotes vegetation in some respects by resulting in the planting of more trees or the promotion of tree protection activities. Through the analysis of the temporal and spatial NDVI, it can be assessed that the vegetation coverage is relatively large and the growth condition of vegetation in this region is good overall.

Keywords: NDVI; vegetation; Savitzky–Golay filtering; spatial and temporal analysis; Ruoergai area

1. Introduction

Global climate change, such as increasing atmospheric temperature, irregular precipitation, and changes in sunshine durations, has been detected, and nowadays the climate change is happening more obviously than the past several decades [1–6]. The changing climate has significantly influenced the growth of vegetation, and the mechanisms of climate change influencing vegetation are complex because vegetation can be influenced by the changing climate and can also adapt to various environments [7–9]. It has been proven that the increasing trend of global warming has influenced vegetation, such as changes in the phenology of vegetation (greening, flowering, and leaf fall) [10–14]. Since the increasing trend will last for the next several decades, it is of vital importance to monitor the changes in vegetation, as vegetation coverage is closely correlated with the gross primary productivity, biosphere, and ecological cycle, which are the dominant factors for balancing ecosystems [15–19]. Thus, timely monitoring of the growth coverage and growth condition of vegetation and acknowledging the temporal and spatial changes in vegetation will not only benefit the development of society but also promote the quality of the ecological environment.

Satellite remote sensing has long been adopted for observing vegetation because it can collect large-area data over long time periods [20–23]. SRS can detect ground objects using the combination of spectral bands of images ranging from the visible to near-infrared and infrared [24]. The vegetation index is commonly designed to maximize the characteristics of vegetations while minimizing other effects, such as soil disturbance and atmospheric effects [25,26]. The vegetation index is a simple and efficient indicator for assessing and evaluating vegetation coverage, biomass, and soil background on the ground [27–31]. The normalized difference vegetation index (NDVI) is calculated as a ratio between the red and near infrared values in traditional fashion: $(NIR - R)/(NIR + R)$. The NDVI is used to estimate the density of green on an area of land, and it is closely related to green vegetation biomass, vegetation growth status, and vegetation photosynthetic capacity [32–35]. NDVI reflects the background influence of the plant canopy, and it is widely used for the retrieval of information regarding vegetation physical parameters and surface vegetation coverages [35,36]. Long time series of NDVI datasets have been used to explore global and regional environmental changes, such as dynamic changes in vegetation and land cover changes. However, the time series of the NDVI dataset contains considerable noise ascribed to atmospheric conditions, such as cloud and aerosol scatter. Additionally, the current Global Inventory Modeling and Mapping Studies (GIMMS) NDVI from Advanced Very High Resolution Radiometer (AVHRR) sensors (National Aeronautics and Space Administration, Washington DC, USA) is a NDVI product available for the period spanning from 1981 to 2006, and the Moderate-resolution Imaging Spectroradiometer (MODIS) NDVI ranges from 2000 to present. Thus, the usage of a single NDVI dataset is not enough if there is a need to explore the temporal changes of a relatively long period [37–39]. Therefore, it is necessary to determine an appropriate method of fusing and generating the standard sequence of NDVI datasets based on different sensors. Further analysis should be conducted based on the rebuilt NDVI sequence.

Several studies have obtained long time series of NDVI datasets over 30 continuous years. However, most of the current studies are based on the comprehensive application of a single dataset, and thus, there is a need to investigate long-term fusions of two or more NDVI datasets when the adoption of a single NDVI dataset can hardly meet the temporal scales (e.g., the NDVI of 1981 to present is needed). There have been comprehensive applications using different sources of NDVI data, and the analysis and comparison of different time scales have proven that there are close correlation relationships between these NDVI products [40–42]. Therefore, the synthesis of continuous NDVI products is possible by fusing NDVI datasets from multiple sources into a uniform set. The Landsat The Enhanced Thematic Mapper Plus NDVI and NDVI calculated by Moderate-resolution Imaging Spectroradiometer (MODIS) products were normalized and compared, and the results show that the NDVI from MODIS and its combination with different sources can be adopted for time-series analysis [43]. Kevin et al. analyzed the difference between the NDVI from MODIS and the AVHRR sensors on the National Oceanic and Atmospheric Administration (NOAA) satellites and established a good relationship model using the two data sources. The different vegetation types using the two sensors from MODIS and NOAA in the United States for the same time period were compared, and the results showed that the synthetic 16-day data were similar and showed good linear relationships [44]. Michele et al. compared the two instruments from the point of view of the user interested in operational crop monitoring using PROBA-V (QinetiQ Space Belgium, Paris, France) instead of SPOT (Spot Image, Toulouse, France), and the results showed a high agreement between these two instruments [45]. Caleb et al. conducted a methodology based on a dynamic framework that was proposed to incorporate additional sources of information into the NDVI time series of agricultural observations for the estimation of phenology [46]. Mao et al. constructed the yearly maximum GIMMS NDVI sequence of the AVHRR and MODIS in Northeast China from 1982 to 2009 using a per-pixel unary linear regression model [47]. Thus, it is possible to obtain a continuous NDVI product with high consistency by fusing the data from a similar remote sensing system. A reconstructed long time series of NDVI datasets can be used to monitor the dynamic changes in vegetation and land transfer types by revealing important information and knowledge. To date, only a few studies have focused on the reconstruction of NDVI time series using two or more NDVI datasets that cover the whole Ruoergai wetland area (RWA). The NDVIs from AVHRR, MODIS, and SPOT are available for the RWA, and they are important for specific vegetation types, vegetation growth characteristics, topography, climate, and other factors. They are all suitable for monitoring the dynamic changes in vegetation in the RWA. The multiple sources of NDVI can reflect the characteristics of vegetation changes, and their respective characteristics can be investigated and obtained to perform efficient data fusion.

To the best of our knowledge, the influencing factors that characterize long-term vegetation changes have not been well explored or investigated in the RWA. The ecological environment of the RWA is vulnerable to the changing climate, as it belongs to high-altitude areas. Meanwhile, the single dataset of NDVI can hardly support the analysis of vegetation in the RWA from 1982 to present. Thus, further in-depth and extensive research is urgently needed to analyze the change and trend of the NDVI in the RWA. Two different NDVI datasets: GIMMS NDVI (1982 to 2006) and MODIS NDVI (2000 to 2018) have been adopted for analyzing the spatial and temporal changes of vegetation in the RWA. In this study, the objectives were to try (1) to apply linear regression analysis of two NDVI datasets between GIMMS and MODIS NDVI of the RWA; (2) to reconstruct the NDVI sequence through the Savitzky–Golay (SG) filtering method using the GIMMS NDVI and MODIS NDVI in RWA; and (3) to analyze the spatial, temporal, and seasonal changes with the trends of vegetations in RWA at different counties.

2. Material and Methods

2.1. Study Area

The RWA is in the northern part of Sichuan Province, which is on the edge of the Qinghai–Tibet Plateau (Figure 1). The south (north) is high (low) in altitude, ranging from 2500 to 5000 m [48]. The climate in this area is a humid monsoon, as it is in the plateau cold zone, with cold winters and cool summers, along with abundant solar radiation, rain, and hot temperatures. The annual rainfall is between 500 and 600 mm/year, with an annual average temperature of 1.4 °C. The average annual sunshine hours is 2389, and the annual average evaporation is 1232 mm [49]. The maximum wind force is 11 m/second, and the wind direction is mostly a northwest wind. Freezing starts in late September each year, with a maximum frozen soil depth of 72 cm, and can only be completely thawed in mid-May [50]. The vegetation in this region is dominated by alpine meadows and marsh vegetation, with a total wetland area of approximately 53 million square kilometers. There are five counties in this area, namely, Aba County, Hongyuan County, Ruoergai County, Luqu County, and Maqu County.

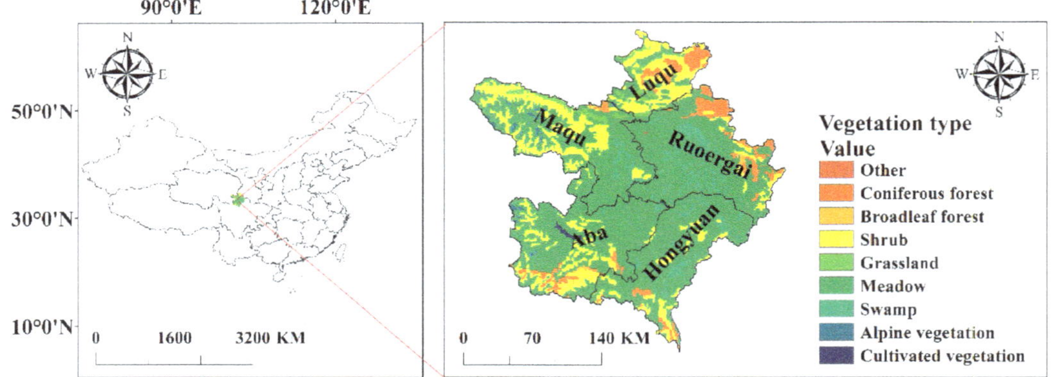

Figure 1. The location of the study area in China and the vegetation types of the Ruoergai wetland area (RWA).

Since 1970, the ecological and environmental problems in the RWA have become increasingly serious. Thus, the RWA was established as a nature reserve in 1994, and a series of ecological restoration measures have been implemented in order to improve the growth of vegetation and increase vegetation coverage [51]. In general, the vegetation growth conditions and evolution trends in the RWA are of great significance to local tourism, animal husbandry, transportation, urban development, and even the sustainable economic and ecological development of the entire southwestern region. Therefore, the vegetation change in the RWA deserves substantial attention.

2.2. Data Source

2.2.1. NDVI Data Source

The remote sensing-obtained NDVI data used in this study are from the GIMMS and MODIS (Santa Barbara Remote Sensing, NASA, USA) NDVI. The GIMMS NDVI can be assessed from the Heihe Project Data Management Center and the Cold and Arid Regions Science Data Center (http://westdc.westgis.ac.cn). The data have been corrected through radiometric correction and cloud removal for quality control. The international maximum value synthesis method is adopted to obtain the NDVI data every half month, and the data spatial resolution is 8 km. The MODIS remote sensing data were downloaded from the vegetation index data (MOD13A3) provided by the National Space Administration (NASA), of which the time resolution was 16 days and the spatial resolution was 1 km (https://ladsweb.modaps.eosdis.nasa.gov/missions-and-measurements/products/MOD13A3/). The data have undergone preprocessing such as the elimination of ground reflections and noise.

2.2.2. Vegetation Types

The vector data of this region were obtained from the National Geographic Information Resource Directory Service System (http://www.webmap.cn/commres.do?method=result100W). The vegetation type data source was from the project "1:1,000,000 China Vegetation Atlas" compiled by the Resource and Environmental Science Data Center of the Chinese Academy of Sciences (CAS) (http://www.resdc.cn). The vegetation types of this region contain nine main types of cultivated vegetation, namely, broad-leaved forest, swamp, grassland, shrub, meadow, alpine vegetation, and coniferous forest. To match the NDVI data, the resolution of the vegetation type data was resampled in ARCGIS (version 10.4, Esri, USA) (https://www.arcgis.com/index.html), and then mask extraction was performed to obtain the vegetation type of this region.

2.3. Method

2.3.1. The Mean Method and Resampling Method

The average annual vegetation NDVI and seasonal average vegetation NDVI in the RWA from 2000 to 2015 were calculated by means of the monthly NDVI value. The formula is defined as follows:

$$\overline{NDVI_x} = \frac{\sum_{i=1}^{12} NDVI_i}{12} \tag{1}$$

In the equation, the variable i varies from 1 to 12, $\overline{NDVI_x}$ is the average annual NDVI, $NDVI_i$ is the NDVI of the i-th month, and i is the month number. The seasonal averages are defined as spring (from March to May), summer (from June to August), autumn (from September to November), and winter (from December to February).

Since the spatial resolution of GIMMS and MODIS NDVI was different, the resampling operation should be performed to achieve consistency. Therefore, the GIMMS NDVI was sampled as 1 km in ArcGIS. Since the MODIS NDVI and the GIMMS NDVI have the same temporal periods from 2000 to 2006, the multiyear average value of seven years was calculated for each pixel to achieve continuity between the two datasets from January to December. The linear fitting formula for each month is defined as follows:

$$NDVI_G = a \times (NDVI_M) + b \tag{2}$$

where $NDVI_G$ is the value of GIMMS NDVI and $NDVI_M$ is the MODIS NDVI. In addition, a is the coefficient, and b is the constant.

2.3.2. Savitzky–Golay Method

It is necessary to determine certain filtering parameters, then use polynomials, and finally realize the filtering method of least square fitting. This method was proposed by Savitzky and Golay in 1964 and improved by Chen Equal in 2004 to make it more suitable for fitting vegetation growth curves [52]. The formula is defined as follows:

$$Y_j^* = \frac{\sum_{i=-m}^{i=m} C_i Y_j + i}{N} \tag{3}$$

where Y_j is the initial NDVI; Y_j^* is the fitted NDVI value; m represents the size of the sliding window; C is the filter coefficient of the i-th NDVI value; N represents the length of the filter, meaning the width of the sliding array is $(2 \times m + 1)$; and j represents the coefficient of the initial NDVI array.

2.3.3. Workflow of Data Processing

The whole workflow of data processing can be divided into four parts (Figure 2). First, the NDVIs from GIMMS and MODIS were preprocessed to achieve the same format, projection, and coordinate system, which are necessary for information extraction. To achieve this, the MODIS Reprojection Tool (MRT) tool was adopted for preprocessing, such as data cropping and projection coordinate conversion of the MODIS NDVI data to ensure a coordinate system consistent with that of the GIMMS NDVI. The vegetation type data and administrative division vector data were used as masks to extract vectors from the RWA. Second, the GIMMS NDVI was resampled to the same resolution as MODIS, and SG was adopted to reconstruct the NDVI time-series data, which aims to improve the quality of the data by removing noise, clouds, and aerosols. Third, the unary linear regression method was used for reconstructing the long-term NDVI sequence in the RWA. For vegetation type data and administrative division data, the collected vegetation type data were combined with the administrative division data to extract the vegetation type data in the RWA. Thus, the map of vegetation types in the RWA was constructed. At the time intersection of the two NDVI datasets from 2000 to 2006, mean value synthesis was performed to obtain the year-by-year NDVI of the RWA from 1982 to 2018, and unary linear regression was used to unify the two datasets. Finally, according to the unified dataset, from the seasonal changes of spring, summer, autumn, and winter and the different time scales of the annual change, the characteristics of the time change and spatial differentiation of the study area were analyzed. The results of the RWA were analyzed through the following four aspects:

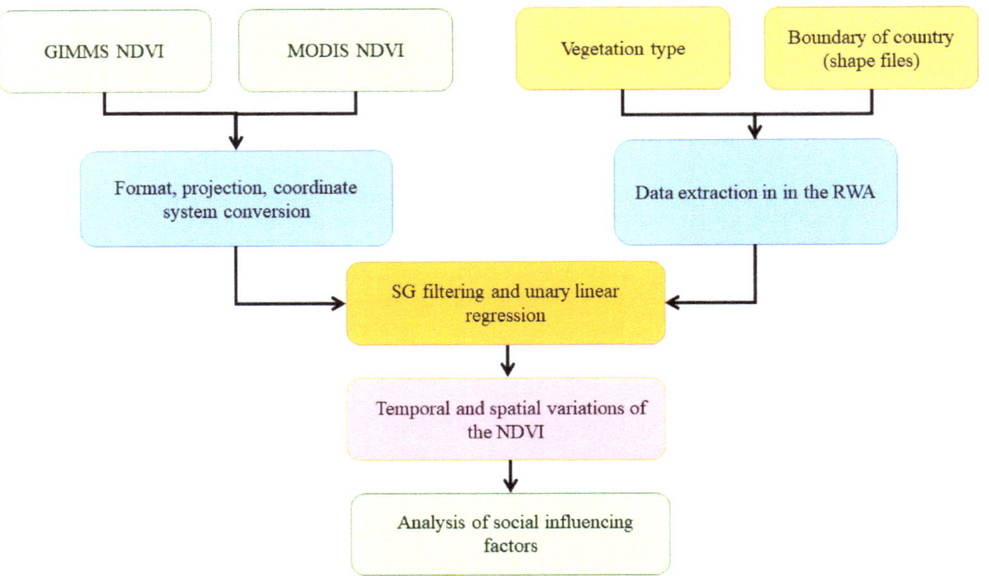

Figure 2. Workflow of data processing. GIMMS: Global Inventory Modeling and Mapping Studies; NDVI: normalized difference vegetation index; MODIS: Moderate-resolution Imaging Spectroradiometer; SG: Savitzky–Golay.

(1) The NDVI data were first preprocessed, and for the results of the filtering process, the GIMMS NDVI was processed using the SG filtering method. The results of the same time period were compared with the NDVI of MODIS to ensure that the accuracy of the filtered data was improved.

(2) The vegetation type, shape file, and NDVI data were all processed and extracted for the RWA in five counties: Aba County, Hongyuan County, Ruoergai County, Luqu County, and Maqu County.

(3) The temporal change in vegetation coverage using different scales, including the four seasons and interannual changes in the NDVI of the RWA was analyzed. Spring was defined as being from March to May, summer was defined as being from June to August, autumn was defined as being from September to November, and winter was defined as being from December to February.

(4) The spatial change in vegetation coverage, including the spatial distribution and change trend of its NDVI value in different seasons and at different scales, was evaluated; additionally, the four seasons and the whole year for the different districts and counties in the RWA were analyzed.

(5) The contributions of social influencing factors were analyzed. The gross domestic product (GDP) was obtained from the National Bureau of Statistics (https://data.stats. gov.cn/), and the detailed data was recorded in country, year, GDP (ten thousand CNY) forms in an independent file. These data were not all available for all counties, and thus, only counties with GDP data were further selected for linear regression. The linear relationships were built between GDP and the NDVI values and the coefficients were obtained.

3. Results

3.1. Preprocessing of the NDVI

The long time series of the GIMMS and MODIS NDVI of the RWA are shown and compared (Figure 3a,b). The data were resampled into monthly values to better match the consistency of the two different sensors that provided the NDVI sequence. The GIMMS NDVI before and after the filtering process is also shown and compared with the MODIS NDVI from February 2002 to September 2006 (Figure 3c,d).

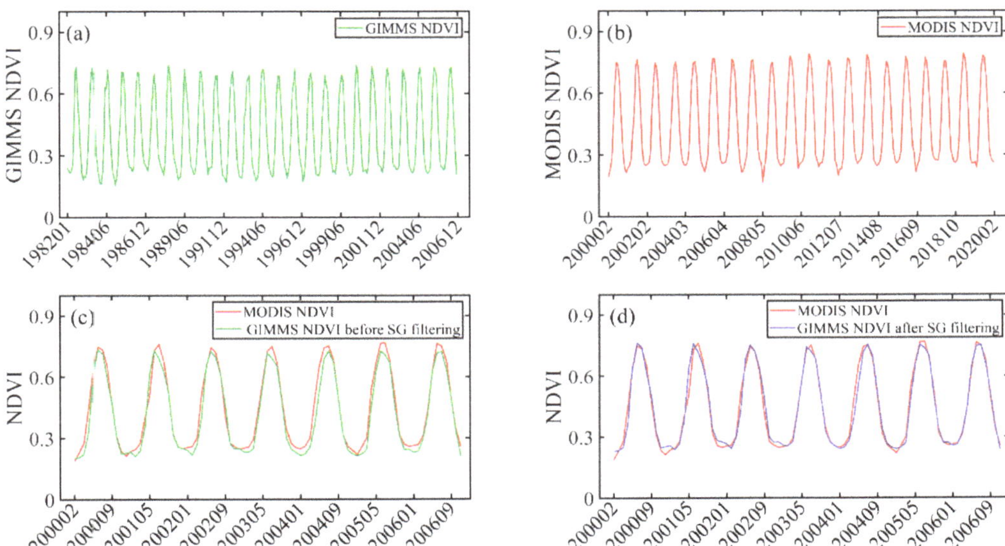

Figure 3. Original sequence of the GIMMS and MODIS NDVI and comparison of the GIMMS NDVI before and after the SG filtering correction. Note: (**a**) represents the original sequence of the GIMMS NDVI; (**b**) represents the original sequence of the MODIS NDVI; (**c**) is the comparison of the GIMMS NDVI before the SG filtering correction and the MODIS NDVI; (**d**) is the comparison of the GIMMS NDVI after the SG filtering correction and the MODIS NDVI. The axis of each subfigure represents the year with the corresponding month.

It can be acknowledged that the curve of GIMMS before SG filtering had poor continuity and prominent abnormal values compared with MODIS. To be more specific, the GIMMS NDVI values before the filtering process were all lower at the peak and bottom than those of MODIS NDVI, and these NDVI values meant that the GIMMS NDVI was not well corrected compared with MODIS NDVI. When the GIMMS NDVI was filtered using the SG method, the data values of GIMMS were similar to those of MODIS. Thus, the current NDVI sequence of GIMIMS for the entire time period (January 1982 to December 2006) should be filtered to remove the effects ascribed from different sensors using the SG filtering method. Thus, the long time series of the NDVI sequence was obtained via the above filtering method (1982–2020).

3.2. Temporal Changes of the NDVI in the RWA

The monthly average NDVI values and the seasonal average NDVI values of the RWA were calculated using the reconstructed NDVI sequence from 1982–2020. The vegetation of this region was analyzed using the NDVI sequence for spring, summer, autumn, and winter (Figure 4). The coefficients of determination and significance values were calculated using the inbuilt function *regress* in MATLAB (version 2019b, University of New Mexico, Albuquerque, USA). It can be directly observed that the NDVI values significantly differed between the four seasons in the RWA, of which the rate of the NDVI in summer increased the fastest and the rate of the NDVI in autumn increased the slowest, as the calculated coefficients of determination of these two seasons were the highest and lowest among the four seasons.

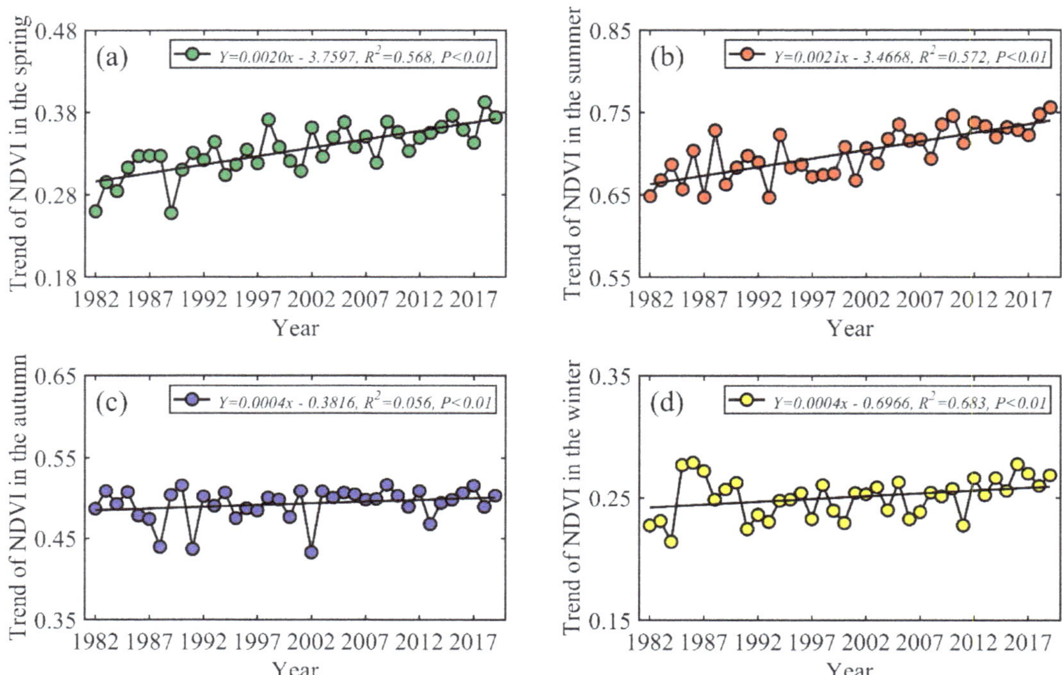

Figure 4. Seasonal trends of the NDVI in the RWA from 1982–2018. Note: (**a–d**) are the NDVI trends for spring, summer, autumn, and winter, respectively. The R^2 and P in the legend of each subplot are the coefficient of determination and significance values, respectively.

The linear regression models were built between time and the NDVI sequence and the ratios were 0.020, 0.021, 0.004, and 0.004/10a and the R^2 were 0.568, 0.572, 0.056, and 0.683 for spring, summer, autumn, and winter, respectively. The vegetation trends in spring and summer were maybe greater than those in autumn and winter. Based on the interannual variation characteristics of the seasonal average NDVI, the seasonal variation trends of the NDVI in the RWA were different from 1982 to 2019. The seasonal changes in the NDVI value in spring decreased dramatically in 1989. The minimum, maximum, and mean values of the NDVI were 0.258, 0.392, and 0.343 for spring, respectively. The growth trends of the NDVI varied significantly between approximately 1988 and 1994, which may be correlated with the changes in climate conditions. After 2000, although there was still an alternating trend of rising and falling, the overall summer NDVI trend was relatively good. The minimum, maximum, and mean values of the NDVI were 0.647, 0.756, and 0.721 in summer, respectively. In autumn, there were three decreasing points (1988, 1991, and 2002), and the corresponding NDVI values were 0.440, 0.437, and 0.433, respectively. The overall change trend of the NDVI in autumn was not obvious. The minimum, maximum, and mean values of the NDVI were 0.433, 0.517, and 0.506 in autumn, respectively. The increasing trend of the NDVI in winter was slightly higher than that in autumn, and the minimum, maximum, and mean values of the NDVI were 0.214, 0.258, and 0.279, respectively.

The annual average NDVI values from 1982 to 2018 were calculated to explore the overall change trend of vegetation in the RWA (Figure 5). The curve of the NDVI sequence was volatile from 1982 to 1998, of which the NDVI increased and decreased one year later. The average NDVI decreased first from 1998, reached its lowest value in 1999, and then increased until 2005. The NDVI values varied between 0.405 and 0.472, and the overall trend of the NDVI sequence in the RWA increased. The average NDVI value was 0.457, the rate of the linear trend of NDVI was 0.013/10a, and the coefficient of determination was 0.683 with a *p* value < 0.01. The minimum annual NDVI value appeared in 2018, with a corresponding value of 0.405, and the maximum annual NDVI value appeared in 1982, with a corresponding value of 0.472.

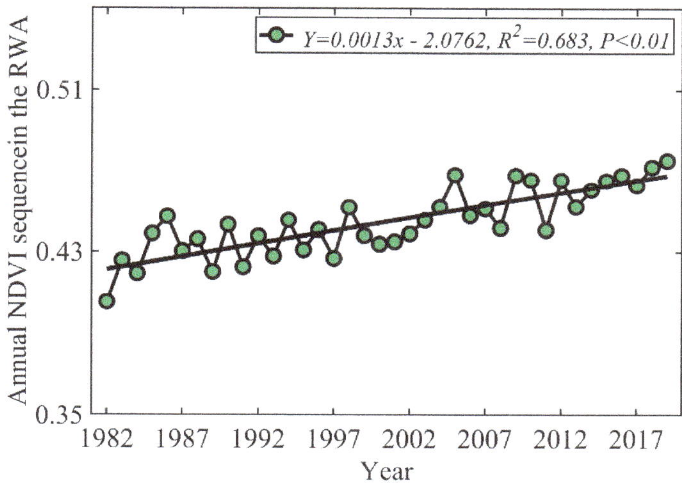

Figure 5. Annual NDVI sequence in the RWA from 1982–2018.

The annual average NDVI value from 1982 to 2004 was lower than that from 1981 to 2017, which indicated that the growth condition of vegetation was not good enough. However, most of the NDVI values were higher than the annual average value after 2004.

3.3. Spatial Changes of the NDVI in the RWA

With the combination of different data sources, the differentiation characteristics of the RWA were assessed and analyzed using long time series of NDVI sequences from different areas. The average NDVI in the RWA from 1982 to 2018 was calculated, and the spatial distribution is shown (Figure 6). The NDVI value intervals were set as 0.000, 0.200, 0.400, 0.600, 0.800, and 1. The NDVI values of 0 indicated that the pixel was water or potential non-vegetation coverage.

The total area in this region was 59,889 pixels, of which the area with values equal to 0 was 2339 pixels, accounting for 0.039% of the area. The area distributed from 0.010 to 0.200 was 18 pixels. The area distributed from 0.210 to 0.400 was 5447 pixels, accounting for 0.091% of the total area. The area distributed at the interval from 0.410 to 0.600 was 51,143 pixels, accounting for 85% of total accounts. The area in the range of NDVI values of 0.610 to 0.800 was 942 pixels, accounting for 0.0158%.

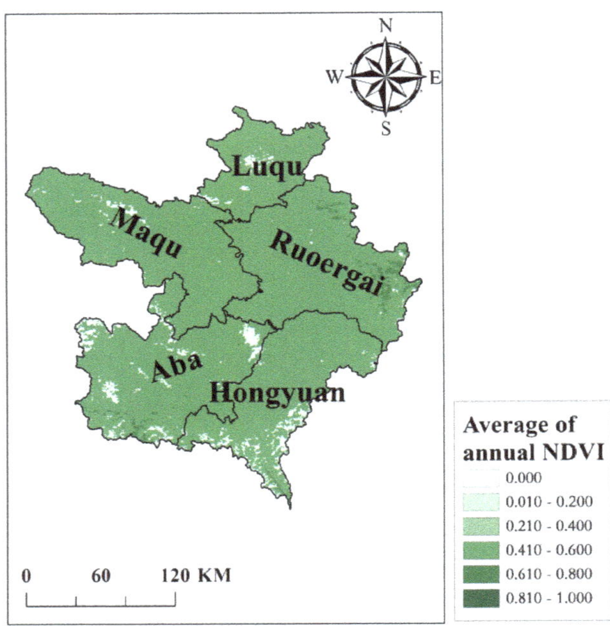

Figure 6. Spatial distribution of the mean NDVI values in the RWA from 1982 to 2018.

The multiyear average of NDVI values ranging from 0.410 to 0.600 occupied most of the area, dominating the vegetation coverage of the RWA. The spatial distribution of the mean NDVI showed that the northeast and southwest had high values. The largest NDVI values were in northeastern Zoige County, southernmost Aba County, and Hongyuan County. The average NDVI sequence larger than 0.600 of the RWA mainly belonged to the vegetation types of coniferous forests and broad-leaved forests. NDVI values of less than 0.200 were mainly located in northwestern Maqu County, northern Aba County, and southeastern Hongyuan County in Gannan Tibetan Autonomous Prefecture and Gansu Province. The reason for this was that the vegetation types distributed in this region were mainly alpine vegetation, swamps, and meadows, and the soil conditions were poor. Through the comprehensive analysis of the temporal and spatial changes in the NDVI sequence, it can be concluded that the vegetation increased.

Considering spatial impacts, the seasonal changes in the NDVI were recalculated for different areas in the RWA. The trends of the maximum and average values of the NDVI in the RWA were analyzed for different seasons and different areas (Figure 7). For seasonal changes in the NDVI, summer had the highest values of NDVI, followed by spring, autumn, and winter. In regard to different regions, the maximum NDVI values of Zoige County were higher than those of the other four counties throughout the year, which indicated the potential good growth of vegetation and good vegetation coverage. The maximum NDVI values of Maqu County were lower than those of the other four counties, and the overall NDVI maximum value was the lowest, which indicated relatively low vegetation coverage. The variation curves of the maximum NDVI of the five counties were consistent. The distribution of maximum NDVI values in different counties was in the following order: Zoige County, Aba County, Hongyuan County, Luqu County, and Maqu County. In regard to the average values of the NDVI, Zoige County had the highest and Aba County had the lowest values. However, the distribution curves of all of the counties were very close. The NDVI average distribution curves of Aba County, Hongyuan County, Maqu County, and

Luqu County in different quarters were relatively close, and the NDVI showed a certain degree of overlap.

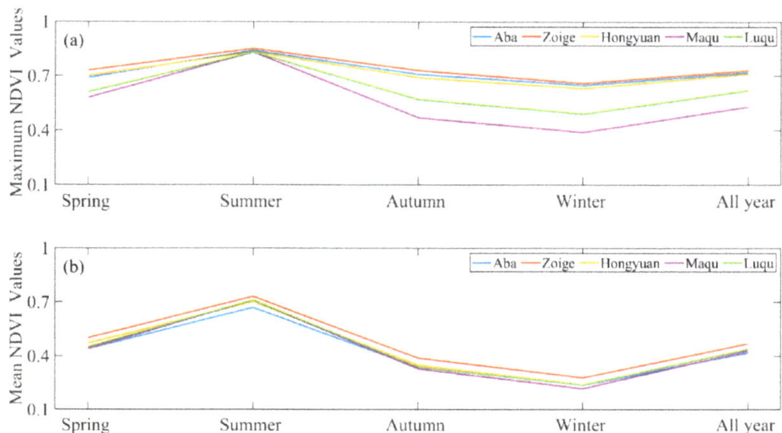

Figure 7. Trends of the maximum and mean values of the NDVI values in different districts and counties of the RWA. Note: (**a**,**b**) indicate the maximum and mean NDVI values, respectively.

The spatial distribution of the mean NDVI values in spring, summer, autumn, and winter in the RWA from 1982 to 2018 are shown in Figure 8. The spatial distributions of the NDVI values in autumn and winter were relatively similar, and the results in spring were similar to those in summer. Most of the NDVI values in the spring were between 0.410 and 0.600 (51,402 pixels), which accounted for 86% of the total number of pixels. The number of pixels of NDVI values ranging from 0.010 to 0.200 and from 0.210 to 0.400 was 28 and 5102, respectively, and the total percentage was 9%. The corresponding number of pixels whose values varied from 0.61 to 0.80 was 1026, which accounted for 2%. It should be acknowledged that the low-value areas of NDVI were mainly distributed in Maqu County and in western Aba County. In contrast, the high-value areas of NDVI were concentrated in eastern Zoige County. The spatial distribution of NDVI values was relatively discrete in summer, with values ranging from 0.610 to 0.800, accounting for 88% of the total pixels. High NDVI values appeared in eastern Maqu County, Aba County, and eastern Zoige County. The land usage classification showed that those regions were mostly covered by vegetation types such as coniferous forests, shrubs, and broad-leaved forests, which improved the vegetation coverage. On the other hand, the low values were in Hongyuan County and Aba County. This result may be ascribed to the reason that this region was closer to the Chengdu Plain and the altitude was relatively low.

The NDVI had a significant decreasing trend in autumn, the NDVI values were most widely distributed between 0.210 and 0.400, and the number of pixels was 48,976, accounting for 82% of the total number of pixels. The main reason was that the NDVI values were influenced by the harvest of crops such as cereal crops, and thus, the vegetation coverage rate decreased. The numbers of pixels with NDVI values ranging from 0.010 to 0.200 and from 0.410 to 0.600 were 508 and 7718, respectively. The corresponding areas with NDVI values concentrated from 0.410 to 0.600 were distributed in the south-central part of Aba County, the southern part of Hongyuan County, and the northeastern part of Zoige County. This result was due to the land being covered by coniferous forest and broad-leaved forest. The number of pixels with values ranging from 0.610 to 0.800 was 341. The spatial distribution of the NDVI values in winter was similar to that in autumn. The difference was that the NDVI values in winter were the lowest, and the NDVI value in winter was the most widely distributed between 0.210 and 0.400, accounting for 85% of the total number of pixels (50,875 pixels). The average value was 0.260, and the number of pixels of the NDVI values concentrated from 0.010 to 0.200 and from 0.410 to 0.600 were 4551 and 2039, respectively. The number of pixels of NDVI values ranging from 0.610 to 0.800 was 80, and this percentage was the lowest. Due to the decrease in temperature, high altitude, crop dormancy, and reduced photosynthesis in winter, the vegetation types were mainly grasslands and meadows, resulting in a significant decrease in vegetation coverage.

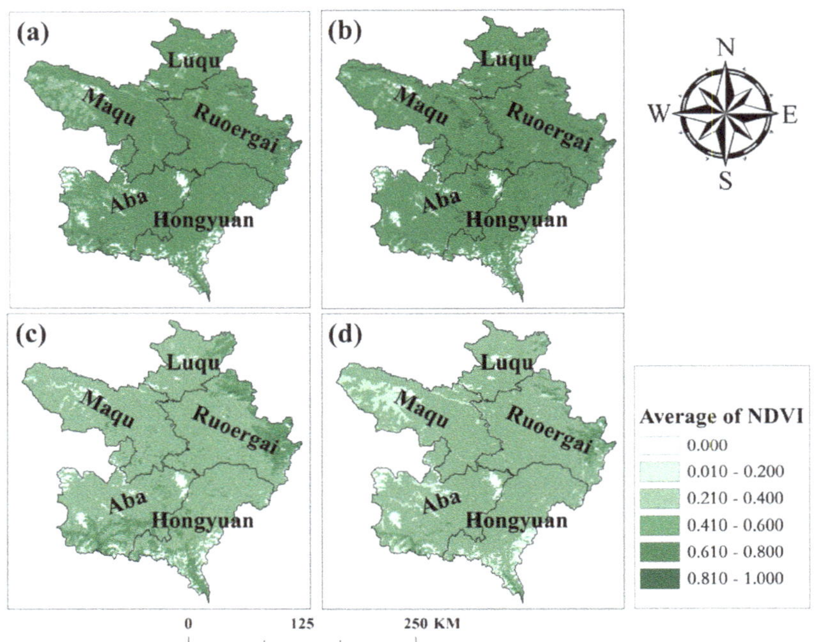

Figure 8. Spatial distribution of the mean NDVI values in different seasons in the RWA from 1982 to 2018: (**a**) spring; (**b**) summer; (**c**) autumn; and (**d**) winter.

In general, the NDVI values in winter, spring, and summer ranged from 0.210 to 0.400, 0.410 to 0.600, and 0.610 to 0.80, respectively (Figure 8). The RWA showed a more obvious trend of improvement in vegetation coverage, and the NDVI values increased. The spatial distribution of NDVI values in autumn was similar to that in winter, with the NDVI values ranging from 0.210 to 0.400. This result was ascribed to the changing characteristics of vegetation in autumn. The photosynthesis of vegetation decreased, and the growth of vegetation was slow, which further resulted in the low NDVI values in that period.

3.4. Analysis of the Contributions of Social Influencing Factors to Vegetation

To explore the contributions of social influencing factors on vegetation, long time series of GDP data were adopted for data analysis. The GDP data of Aba, Ruoergai, and Hongyuan were available, whereas the data of the other counties were not found. The linear regression was performed with the time and NDVI sequence, indicating that as time progresses, NDVI presents an increasing trend. Similar relationships were found between GDP and time. Then the linear regression model was performed independently between the GDP and NDVI sequence where the GDP data was available. The linear trend coefficients, regression equations, coefficient of determination (R^2), and significant values were calculated, as shown in Figure 9.

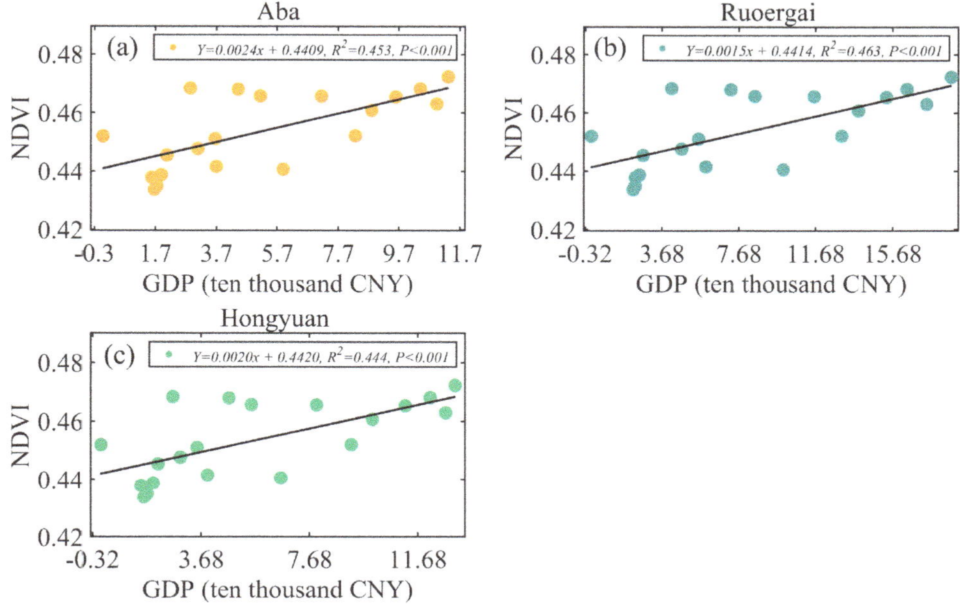

Figure 9. Linear regression analysis between the GDP and NDVI at the county level. Note: (**a**) Aba; (**b**) Ruoergai; and (**c**) Hongyuan.

The ratios of the regression models were all positive, indicating that GDP had a positive relationship with NDVI, of which the ratios were 0.0024, 0.0015, and 0.0020 for Aba, Ruoergai, and Hongyuan, respectively. Thus, the contribution of GDP to vegetation was observed to be in the following order: Aba, Hongyuan, and Ruoergai. The R^2 values were 0.453, 0.463, and 0.444, respectively, with significant values lower than 0.001. Since GDP was positively correlated with NDVI and the p values were all significant, this implies that the good growth of vegetation in these regions may be influenced by socioeconomic-related factors, such as the planting of more trees and tree protection activities.

4. Discussion

The long time series NDVI sequence of the RWA was reconstructed using the NDVI from GIMMS and MODIS. The rebuilding was performed using only two NDVI sources; thus, the fusion of more data sources, such as the NDVI calculated from SPOT, GF-1, GF-2, and TM/ETM, needs to be investigated in order to evaluate the proposed method in future analyses. The fusion of data from multiple sources may improve the quality of the NDVI sequences; if so, the quantitative analysis will be more reliable [53–55]. However, although we tried to exclude potential uncertainties, there remained three main uncertainties in this study. First, the SG filtering method was applied to exclude the uncertainties from the NDVI sequence; however, noise may still remain [56,57]. The uncertainty from noise, such as clouds and aerosols, may influence the results. Therefore, more filtering methods, such as logistic filtering, wavelet filtering, Gaussian filtering, and Fourier harmonic timing analysis method filtering, could be adopted to compare the filtering results [58,59]. Ultimately, a combination of filtering methods may effectively improve the quality of the NDVI sequence and reduce noise. Second, the spatial resolution of GIMMS NDVI was resampled from 8 km to 1 km, thus, there are uncertainties that may be generated during the process of resampling using the interpolation method [60,61]. Third, meteorological information was not implemented to analyze the dynamic changes in vegetation in the RWA; it is known that climatic variables, such as temperature, precipitation, wind speed, and air humidity, are important influencing factors for the growth of vegetation [62–65]. Moreover, terrain data were not considered, and the terrain would profoundly influence the regional climate and have some impact on vegetation. In addition, the phenology of vegetation was another important influencing factor, and its impact should be assessed and evaluated in order to exclude uncertainties in assessing seasonal changes in the NDVI of the RWA in different counties [66,67]. Thus, the combined impacts of climate and phenology on vegetation should be further investigated. Furthermore, policy implications played important roles in determining the NDVI values in the long time-series data. The spatial and temporal NDVI implied that the vegetation in the RWA grew well and that the potential vegetation coverage was relatively large.

5. Conclusions

A long time series of NDVI sequences of the RWA was acquired by filtering the NDVI of GIMMS from 1982 to 2006 and the NDVI of MODIS from 2000 to 2018 using the Savitzky–Golay filtering method. Through the process of linear regression, the spatial scale conversion of the two datasets was realized. The temporal and spatial variations in vegetation were assessed and evaluated using the NDVI sequence of the RWA. The SG filtering method was adopted in order to exclude the noise of the NDVI sequence, producing a smoothed curve with good continuity and eliminating many outliers. The corrected GIMMS NDVI and MODIS NDVI values were close and consistent. The NDVI in the RWA had different seasonal trends with obvious fluctuations to various degrees. The linear trend rate of the seasonal average NDVI value was summer (0.021/10a), followed by spring (0.020/10a), winter (0.005/10a), and autumn (0.004/10a). The overall NDVI values of the RWA varied from 0.405 to 0.472, and the overall trend increased as 0.013/10a. Zoige County had the maximum NDVI value, followed by Aba County, Hongyuan County, Luqu County, and Maqu County. The vegetation variations in the RWA showed overall good development, with the main distribution range of the NDVI values being from 0.410 to 0.600. The seasonal changes and annual variations in the NDVI of the RWA may imply the potentially good development of vegetation, and the increasing trend of the NDVI sequence may indicate increased vegetation coverage.

Author Contributions: Conceptualization: Y.G., J.Z., W.W., and G.L.; methodology: Y.G., J.Z., and G.L.; software: Y.G., J.Z., W.W., S.H., L.W., and G.L.; validation: Y.G., J.Z., W.W., S.H., and L.W.; formal analysis: Y.G., J.Z., W.W., S.H., L.W., and C.R.B.; investigation: Y.G. and J.Z.; resources: Y.G. and J.Z.; data curation: Y.G. and J.Z.; writing—original draft preparation: Y.G. and J.Z.; writing—review and editing: Y.G., J.Z., and C.R.B.; visualization: Y.G., J.Z., and C.R.B.; supervision: W.W. and C.R.B.; project administration: W.W.; funding acquisition: W.W. All authors have read and agreed to the published version of the manuscript.

Funding: This research was supported by the Strategic Priority Research Program of the Chinese Academy of Sciences (Grant No. XDA20020202, XDA19040101, XDA19040304).

Institutional Review Board Statement: Not applicable.

Informed Consent Statement: Not applicable.

Data Availability Statement: The MODIS NDVI can be acquired at https://modis.gsfc.nasa.gov/data/dataprod/mod13.php, and the GIMMS NDVI can be obtained at https://iridl.ldeo.columbia.edu/SOURCES/.UMD/.GLCF/.GIMMS/.NDVIg/.global/.dataset_documentation.html.

References

1. Hoegh-Guldberg, O.; Jacob, D.; Taylor, M.; Bolaños, T.G.; Bindi, M.; Brown, S.; Camilloni, I.; Diedhiou, A.; Djalante, R.; Ebi, K. The human imperative of stabilizing global climate change at 1.5 C. *Science* **2019**, *365*, eaaw6974. [CrossRef] [PubMed]
2. Seneviratne, S.I.; Donat, M.G.; Mueller, B.; Alexander, L.V. No pause in the increase of hot temperature extremes. *Nat. Clim. Chang.* **2014**, *4*, 161–163. [CrossRef]

3. Gottfried, M.; Pauli, H.; Futschik, A.; Akhalkatsi, M.; Barančok, P.; Alonso, J.L.B.; Coldea, G.; Dick, J.; Erschbamer, B.; Kazakis, G. Continent-wide response of mountain vegetation to climate change. *Nat. Clim. Chang.* **2012**, *2*, 111–115. [CrossRef]

4. Franzke, C.L. Warming trends: Nonlinear climate change. *Nat. Clim. Chang.* **2014**, *4*, 423–424. [CrossRef]

5. Howe, P.D.; Markowitz, E.M.; Lee, T.M.; Ko, C.-Y.; Leiserowitz, A. Global perceptions of local temperature change. *Nat. Clim. Chang.* **2013**, *3*, 352–356. [CrossRef]

6. Mallakpour, I.; Villarini, G. The changing nature of flooding across the central United States. *Nat. Clim. Chang.* **2015**, *5*, 250–254. [CrossRef]

7. Hirabayashi, Y.; Mahendran, R.; Koirala, S.; Konoshima, L.; Yamazaki, D.; Watanabe, S.; Kim, H.; Kanae, S. Global flood risk under climate change. *Nat. Clim. Chang.* **2013**, *3*, 816–821. [CrossRef]

8. Asseng, S.; Ewert, F.; Rosenzweig, C.; Jones, J.W.; Hatfield, J.L.; Ruane, A.C.; Boote, K.J.; Thorburn, P.J.; Rötter, R.P.; Cammarano, D. Uncertainty in simulating wheat yields under climate change. *Nat. Clim. Chang.* **2013**, *3*, 827–832. [CrossRef]

9. Challinor, A.J.; Watson, J.; Lobell, D.B.; Howden, S.; Smith, D.; Chhetri, N. A meta-analysis of crop yield under climate change and adaptation. *Nat. Clim. Chang.* **2014**, *4*, 287–291. [CrossRef]

10. Fu, Y.H.; Zhao, H.; Piao, S.; Peaucelle, M.; Peng, S.; Zhou, G.; Ciais, P.; Huang, M.; Menzel, A.; Peñuelas, J. Declining global warming effects on the phenology of spring leaf unfolding. *Nature* **2015**, *526*, 104–107. [CrossRef]

11. Liu, Q.; Fu, Y.H.; Zhu, Z.; Liu, Y.; Liu, Z.; Huang, M.; Janssens, I.A.; Piao, S. Delayed autumn phenology in the Northern Hemisphere is related to change in both climate and spring phenology. *Glob. Chang. Biol.* **2016**, *22*, 3702–3711. [CrossRef] [PubMed]

12. Vitasse, Y.; Signarbieux, C.; Fu, Y.H. Global warming leads to more uniform spring phenology across elevations. *Proc. Natl. Acad. Sci. USA* **2018**, *115*, 1004–1008. [CrossRef] [PubMed]

13. Khorsand Rosa, R.; Oberbauer, S.F.; Starr, G.; Parker La Puma, I.; Pop, E.; Ahlquist, L.; Baldwin, T. Plant phenological responses to a long-term experimental extension of growing season and soil warming in the tussock tundra of Alaska. *Glob. Chang. Biol.* **2015**, *21*, 4520–4532. [CrossRef] [PubMed]

14. White, M.A.; de Beurs, K.M.; Didan, K.; Inouye, D.W.; Richardson, A.D.; Jensen, O.P.; O'KEEFE, J.; Zhang, G.; Nemani, R.R.; van Leeuwen, W.J. Intercomparison, interpretation, and assessment of spring phenology in North America estimated from remote sensing for 1982–2006. *Glob. Chang. Biol.* **2009**, *15*, 2335–2359. [CrossRef]

15. Zhang, Q.; Cheng, Y.-B.; Lyapustin, A.I.; Wang, Y.; Xiao, X.; Suyker, A.; Verma, S.; Tan, B.; Middleton, E.M. Estimation of crop gross primary production (GPP): I. impact of MODIS observation footprint and impact of vegetation BRDF characteristics. *Agric. For. Meteorol.* **2014**, *191*, 51–63. [CrossRef]

16. Wu, C.; Chen, J.M.; Huang, N. Predicting gross primary production from the enhanced vegetation index and photosynthetically active radiation: Evaluation and calibration. *Remote Sens. Environ.* **2011**, *115*, 3424–3435. [CrossRef]

17. Tang, Z.; Xu, W.; Zhou, G.; Bai, Y.; Li, J.; Tang, X.; Chen, D.; Liu, Q.; Ma, W.; Xiong, G. Patterns of plant carbon, nitrogen, and phosphorus concentration in relation to productivity in China's terrestrial ecosystems. *Proc. Natl. Acad. Sci. USA* **2018**, *115*, 4033–4038. [CrossRef]

18. Li, X.; Liang, S.; Yu, G.; Yuan, W.; Cheng, X.; Xia, J.; Zhao, T.; Feng, J.; Ma, Z.; Ma, M. Estimation of gross primary production over the terrestrial ecosystems in China. *Ecol. Model.* **2013**, *261*, 80–92. [CrossRef]

19. Joiner, J.; Yoshida, Y.; Vasilkov, A.; Schaefer, K.; Jung, M.; Guanter, L.; Zhang, Y.; Garrity, S.; Middleton, E.; Huemmrich, K. The seasonal cycle of satellite chlorophyll fluorescence observations and its relationship to vegetation phenology and ecosystem atmosphere carbon exchange. *Remote Sens. Environ.* **2014**, *152*, 375–391. [CrossRef]

20. Ozesmi, S.L.; Bauer, M.E. Satellite remote sensing of wetlands. *Wetl. Ecol. Manag.* **2002**, *10*, 381–402. [CrossRef]

21. Demarez, V.; Helen, F.; Marais-Sicre, C.; Baup, F. In-season mapping of irrigated crops using Landsat 8 and Sentinel-1 time series. *Remote Sens.* **2019**, *11*, 118. [CrossRef]

22. Gao, Q.; Zribi, M.; Escorihuela, M.J.; Baghdadi, N.; Segui, P.Q. Irrigation mapping using Sentinel-1 time series at field scale. *Remote Sens.* **2018**, *10*, 1495. [CrossRef]

23. Torbick, N.; Chowdhury, D.; Salas, W.; Qi, J. Monitoring rice agriculture across myanmar using time series Sentinel-1 assisted by Landsat-8 and PALSAR-2. *Remote Sens.* **2017**, *9*, 119. [CrossRef]

24. Tsai, Y.-L.S.; Dietz, A.; Oppelt, N.; Kuenzer, C. Wet and dry snow detection using Sentinel-1 SAR data for mountainous areas with a machine learning technique. *Remote Sens.* **2019**, *11*, 895. [CrossRef]

25. Dong, T.; Liu, J.; Shang, J.; Qian, B.; Ma, B.; Kovacs, J.M.; Walters, D.; Jiao, X.; Geng, X.; Shi, Y. Assessment of red-edge vegetation indices for crop leaf area index estimation. *Remote Sens. Environ.* **2019**, *222*, 133–143. [CrossRef]

26. Pôças, I.; Calera, A.; Campos, I.; Cunha, M. Remote sensing for estimating and mapping single and basal crop coefficientes: A review on spectral vegetation indices approaches. *Agric. Water Manag.* **2020**, *233*, 106081. [CrossRef]

27. Suzuki, R.; Tanaka, S.; Yasunari, T. Relationships between meridional profiles of satellite-derived vegetation index (NDVI) and climate over Siberia. *Int. J. Climatol.* **2015**, *20*, 955–967. [CrossRef]

28. Prasad, A.; Singh, R.; Tare, V.; Kafatos, M. Use of vegetation index and meteorological parameters for the prediction of crop yield in India. *Int. J. Remote Sens.* **2007**, *28*, 5207–5235. [CrossRef]

29. Hill, M.J. Vegetation index suites as indicators of vegetation state in grassland and savanna: An analysis with simulated SENTINEL 2 data for a North American transect. *Remote Sens. Environ.* **2013**, *137*, 94–111. [CrossRef]

30. Guo, Y.; Wang, H.; Wu, Z.; Wang, S.; Sun, H.; Senthilnath, J.; Wang, J.; Robin Bryant, C.; Fu, Y. Modified Red Blue Vegetation Index for Chlorophyll Estimation and Yield Prediction of Maize from Visible Images Captured by UAV. *Sensors* **2020**, *20*, 5055. [CrossRef]

31. Guo, Y.; Yin, G.; Sun, H.; Wang, H.; Chen, S.; Senthilnath, J.; Wang, J.; Fu, Y. Scaling Effects on Chlorophyll Content Estimations with RGB Camera Mounted on a UAV Platform Using Machine-Learning Methods. *Sensors* **2020**, *20*, 5130. [CrossRef] [PubMed]

32. Goward, S.N.; Markham, B.; Dye, D.G.; Dulaney, W.; Yang, J. Normalized difference vegetation index measurements from the Advanced Very High Resolution Radiometer. *Remote Sens. Environ.* **1991**, *35*, 257–277. [CrossRef]

33. Zhang, L.; Qiao, N.; Baig, M.H.A.; Huang, C.; Lv, X.; Sun, X.; Zhang, Z. Monitoring vegetation dynamics using the universal normalized vegetation index (UNVI): An optimized vegetation index-VIUPD. *Remote Sens. Lett.* **2019**, *10*, 629–638. [CrossRef]

34. Barbosa, H.A.; Kumar, T.L.; Paredes, F.; Elliott, S.; Ayuga, J. Assessment of Caatinga response to drought using meteosat-SEVIRI normalized difference vegetation index (2008–2016). *ISPRS J. Photogramm. Remote Sens.* **2019**, *148*, 235–252. [CrossRef]

35. Hu, X.; Ren, H.; Tansey, K.; Zheng, Y.; Ghent, D.; Liu, X.; Yan, L. Agricultural drought monitoring using European Space Agency Sentinel 3A land surface temperature and normalized difference vegetation index imageries. *Agric. For. Meteorol.* **2019**, *279*, 107707. [CrossRef]

36. Guha, S.; Govil, H.; Diwan, P. Analytical study of seasonal variability in land surface temperature with normalized difference vegetation index, normalized difference water index, normalized difference built-up index, and normalized multiband drought index. *J. Appl. Remote Sens.* **2019**, *13*, 024518.

37. Fensholt, R.; Proud, S.R. Evaluation of earth observation based global long term vegetation trends—Comparing GIMMS and MODIS global NDVI time series. *Remote Sens. Environ.* **2012**, *119*, 131–147. [CrossRef]

38. Wang, J.; Dong, J.; Liu, J.; Huang, M.; Li, G.; Running, S.W.; Smith, W.K.; Harris, W.; Saigusa, N.; Kondo, H. Comparison of gross primary productivity derived from GIMMS NDVI3g, GIMMS, and MODIS in Southeast Asia. *Remote Sens.* **2014**, *6*, 2108–2133. [CrossRef]

39. Du, J.-Q.; Shu, J.-M.; Wang, Y.-H.; Li, Y.-C.; Zhang, L.-B.; Guo, Y. Comparison of GIMMS and MODIS normalized vegetation index composite data for Qing-Hai-Tibet Plateau. *Ying Yong Sheng Tai Xue Bao J. Appl. Ecol.* **2014**, *25*, 533–544.

40. Geng, L.; Ma, M.; Wang, X.; Yu, W.; Jia, S.; Wang, H. Comparison of Eight Techniques for Reconstructing Multi-Satellite Sensor Time-Series NDVI Data Sets in the Heihe River Basin, China. *Remote Sens.* **2014**, *6*, 2024–2049. [CrossRef]

41. Kiage, L.M.; Nan, D.W. Using NDVI from MODIS to Monitor Duckweed Bloom in Lake Maracaibo, Venezuela. *Water Resour. Manag.* **2009**, *23*, 1125–1135. [CrossRef]

42. Filippa, G.; Cremonese, E.; Migliavacca, M.; Galvagno, M.; Sonnentag, O.; Humphreys, E.; Hufkens, K.; Ryu, Y.; Verfaillie, J.; di Cella, U.; et al. NDVI derived from near-infrared-enabled digital cameras: Applicability across different plant functional types. *Agric. Forest Meteorol.* **2018**, *249*, 275–285. [CrossRef]

43. Wenxia, G.; Huanfeng, S.; Liangpei, Z.; Wei, G. Normalization of NDVI from Different Sensor System using MODIS Products as Reference. *IOP Conf. Ser. Earth Environ. Ence* **2014**, *17*, 012225. [CrossRef]

44. Gallo, K.; Ji, L.; Reed, B.; Eidenshink, J.; Dwyer, J. Multi-platform comparisons of MODIS and AVHRR normalized difference vegetation index data. *Remote Sens. Environ.* **2005**, *99*, 221–231. [CrossRef]

45. Meroni, M.; Fasbender, D.; Balaghi, R.; Dali, M.; Haffani, M.; Haythem, I.; Hooker, J.; Lahlou, M.; Lopez-Lozano, R.; Mahyou, H. Evaluating NDVI data continuity between SPOT-VEGETATION and PROBA-V missions for operational yield forecasting in North African countries. *IEEE Trans. Geosci. Remote Sens.* **2015**, *54*, 795–804. [CrossRef]

46. Bernardis, C.D.; Vicente-Guijalba, F.; Martinez-Marin, T.; Lopez-Sanchez, J.M. Contribution to Real-Time Estimation of Crop Phenological States in a Dynamical Framework Based on NDVI Time Series: Data Fusion with SAR and Temperature. *IEEE J. Sel. Top. Appl. Earth Obs. Remote Sens.* **2016**, *9*, 3512–3523. [CrossRef]

47. Dehua, M.; Zongming, W.; Ling, L.; Guang, Y. Correlation Analysis between NDVI and Climate in Northeast China based on AVHRR and GIMMS Data Sources. *Remote Sens. Technol. Appl.* **2012**, *27*, 81–89.

48. Xiao, D.R.; Tian, B.; Tian, K.; Yang, Y. Landscape patterns and their changes in Sichuan Ruoergai Wetland National Nature Reserve. *Acta Ecol. Sin.* **2010**, *30*, 27–32. [CrossRef]

49. Bian, J.; Li, A.; Deng, W. Estimation and analysis of net primary Productivity of Ruoergai wetland in China for the recent 10 years based on remote sensing. *Procedia Environ. Sci.* **2010**, *2*, 288–301. [CrossRef]

50. Gai, N.; Pan, J.; Tang, H.; Chen, S.; Chen, D.; Zhu, X.; Lu, G.; Yang, Y. Organochlorine pesticides and polychlorinated biphenyls in surface soils from Ruoergai high altitude prairie, east edge of Qinghai-Tibet Plateau. *Sci. Total Environ.* **2014**, *478*, 90–97. [CrossRef]

51. Zhang, X.; Liu, H.; Baker, C.; Graham, S. Restoration approaches used for degraded peatlands in Ruoergai (Zoige), Tibetan Plateau, China, for sustainable land management. *Ecol. Eng.* **2012**, *38*, 86–92. [CrossRef]

52. Atif, A.; Khalid, M. Saviztky–Golay Filtering for Solar Power Smoothing and Ramp Rate Reduction Based on Controlled Battery Energy Storage. *IEEE Access* **2020**, *8*, 33806–33817. [CrossRef]

53. Youzhi, A.N.; Gao, W.; Gao, Z.; Liu, C.; Shi, R. Trend analysis for evaluating the consistency of Terra MODIS and SPOT VGT NDVI time series products in China. *Front. Earth Sci.* **2015**, *9*, 125–136.

54. Raynolds, M.K.; Walker, D.A.; Epstein, H.E.; Pinzon, J.E.; Tucker, C.J. A new estimate of tundra-biome phytomass from trans-Arctic field data and AVHRR NDVI. *Remote Sens. Lett.* **2012**, *3*, 403–411. [CrossRef]

55.   Fensholt, R.; Sandholt, I.; Proud, S.R.; Stisen, S.; Rasmussen, M.O. Assessment of MODIS sun-sensor geometry variations effect on observed NDVI using MSG SEVIRI geostationary data. *Int. J. Remote Sens.* **2010**, *31*, 6163–6187. [CrossRef]
56.   Chen, J.; Jnsson, P.; Tamura, M.; Gu, Z.; Matsushita, B.; Eklundh, L. A simple method for reconstructing a high-quality NDVI time-series data set based on the Savitzky–Golay filter. *Remote Sens. Environ.* **2004**, *91*, 332–344. [CrossRef]
57.   Jinhu, B.; Ainong, L.; Mengqiang, S.; Liqun, M.; Jingang, J. Reconstruction of NDVI time-series datasets of MODIS based on Savitzky-Golay filter. *J. Remote Sens.* **2010**, *14*, 725–741.
58.   Yang, Y.; Luo, J.; Huang, Q.; Wu, W.; Sun, Y. Weighted Double-Logistic Function Fitting Method for Reconstructing the High-Quality Sentinel-2 NDVI Time Series Data Set. *Remote Sens.* **2019**, *11*, 2342. [CrossRef]
59.   Bachoo, A.; Archibald, S. Influence of Using Date-Specific Values when Extracting Phenological Metrics from 8-day Composite NDVI Data, Analysis of Multi-temporal Remote Sensing Images, 2007. In Proceedings of the 2007 International Workshop on the Analysis of Multi-temporal Remote Sensing Images (MultiTemp 2007), Leuven, Belgium, 18–20 July 2007.
60.   Herrera, S.; Kotlarski, S.; Soares, P.M.; Cardoso, R.M.; Jaczewski, A.; Gutiérrez, J.M.; Maraun, D. Uncertainty in gridded precipitation products: Influence of station density, interpolation method and grid resolution. *Int. J. Climatol.* **2019**, *39*, 3717–3729. [CrossRef]
61.   Xie, Y.; Chen, T.-B.; Lei, M.; Yang, J.; Guo, Q.-J.; Song, B.; Zhou, X.-Y. Spatial distribution of soil heavy metal pollution estimated by different interpolation methods: Accuracy and uncertainty analysis. *Chemosphere* **2011**, *82*, 468–476. [CrossRef]
62.   Bhatt, U.S.; Walker, D.A.; Raynolds, M.K.; Bieniek, P.A.; Zhang, J. Changing seasonality of panarctic tundra vegetation in relationship to climatic variables. *Environ. Res. Lett.* **2017**, *12*, 055003. [CrossRef]
63.   Liu, Q.; Yang. Z.; Cui, B.; Sun, T. Temporal trends of hydro-climatic variables and runoff response to climatic variability and vegetation changes in the Yiluo River basin, China. *Hydrol. Process.* **2010**, *23*, 3030–3039. [CrossRef]
64.   Ichii, K.; Kawabata, A.; Yamaguchi, Y. Global correlation analysis for NDVI and climatic variables and NDVI trends: 1982–1990. *Int. J. Remote Sens.* **2002**, *23*, 3873–3878. [CrossRef]
65.   Liu, Y.; Li, L.-H.; Chen, X.; Zhang, R.; Yang, J.-M. Temporal-spatial variations and influencing factors of vegetation cover in Xinjiang from 1982 to 2013 based on GIMMS-NDVI3g. *Glob. Planet. Chang.* **2018**, *169*, 145–155. [CrossRef]
66.   Zhang, X.; Friedl, M.A.; Schaaf, C.B.; Strahler, A.H.; Hodges, J.C.F.; Gao, F.; Reed, B.C.; Huete, A. Monitoring vegetation phenology using MODIS. *Remote Sens. Environ.* **2003**, *84*, 471–475. [CrossRef]
67.   Piao, S.; Fang, J.; Zhou, L.; Ciais, P.; Zhu, B. Variations in satellite-derived phenology in China's temperate vegetation. *Glob. Chang. Biol.* **2010**, *12*, 672–685. [CrossRef]

*forests*

Article

Monitoring Mangrove Forest Degradation and Regeneration: Landsat Time Series Analysis of Moisture and Vegetation Indices at Rabigh Lagoon, Red Sea

Mohammed Othman Aljahdali [1,*], Sana Munawar [2] and Waseem Razzaq Khan [3,4,*]

[1] Department of Biological Sciences, Faculty of Science, King Abdulaziz University (KAU), P.O. Box 80203, Jeddah 21589, Saudi Arabia
[2] Umweltfernerkundung und Geoinformatik, Universität Trier, 54296 Trier, Germany; s6ssmuna@uni-trier.de
[3] Department of Forestry Science and Biodiversity, Faculty of Forestry and Environment, Universiti Putra Malaysia, Sri Serdang, Selangor 43300, Malaysia
[4] Institut Ekosains Borneo, Universiti Putra Malaysia Kampus Bintulu, Sarawak 97008, Malaysia
* Correspondence: moaljahdali@kau.edu.sa (M.O.A.); khanwaseem@upm.edu.my (W.R.K.)

Abstract: Rabigh Lagoon, located on the eastern coast of the Red Sea, is an ecologically rich zone in Saudi Arabia, providing habitat to *Avicennia marina* mangrove trees. The environmental quality of the lagoon has been decaying since the 1990s mainly from sedimentation, road construction, and camel grazing. However, because of remedial measures, the mangrove communities have shown some degree of restoration. This study aims to monitor mangrove health of Rabigh Lagoon during the time it was under stress from road construction and after the road was demolished. For this purpose, time series of EVI (Enhanced Vegetation Index), MSAVI (Modified, Soil-Adjusted Vegetation Index), NDVI (Normalized Difference Vegetation Index), and NDMI (Normalized Difference Moisture Index) have been used as a proxy to plant biomass and indicator of forest disturbance and recovery. Long-term trend patterns, through linear, least square regression, were estimated using 30 m annual Landsat surface-reflectance-derived indices from 1986 to 2019. The outcome of this study showed (1) a positive trend over most of the study region during the evaluation period; (2) most trend slopes were gradual and weakly positive, implying subtle changes as opposed to abrupt changes; (3) all four indices divided the times series into three phases: degraded mangroves, slow recovery, and regenerated mangroves; (4) MSAVI performed best in capturing various trend patterns related to the greenness of vegetation; and (5) NDMI better identified forest disturbance and recovery in terms of water stress. Validating observed patterns using only the regression slope proved to be a challenge. Therefore, water quality parameters such as salinity, pH/dissolved oxygen should also be investigated to explain the calculated trends.

Keywords: forest change; degradation; regeneration; geospatial–temporal analysis; trend

Citation: Aljahdali, M.O.; Munawar, S.; Khan, W.R. Monitoring Mangrove Forest Degradation and Regeneration: Landsat Time Series Analysis of Moisture and Vegetation Indices at Rabigh Lagoon, Red Sea. *Forests* **2021**, *12*, 52. https://doi.org/10.3390/f12010052

Received: 15 October 2020
Accepted: 15 December 2020
Published: 1 January 2021

Publisher's Note: MDPI stays neutral with regard to jurisdictional claims in published maps and institutional affiliations.

1. Introduction

Mangrove swamps are a common feature in coastal areas that are exposed to the daily fluctuation of tides. They are predominantly found in muddy substrate with low wave energy and hypersaline environment [1–7]. Mangroves serve as an important habitat for fish communities and other benthic flora and fauna [8]. Moreover, they play a key role in shoreline protection and waste assimilation through the purification of marine water and surrounding air. Mangroves are also important for carbon sequestration since they are a major sink for carbon by storing it in the sediments [9,10]. In addition, mangrove trees have shown great tolerance to low levels of dissolved oxygen in coastal environments [11–14].

The spatial distribution of mangrove trees along the Red Sea coast of Saudi Arabia was found in 104 locations with an estimated total area of 3452 hectares (ha) [15]. Two major species common to this area are *Avicennia marina* and *Rhizophora mucronata*. Lagoons share 15% of the globes' coastal areas. Mangrove swamps of Red Sea are essential for various ecological functions such as nursery for commercial fish species [16], protection of coral reefs [17], and provision of nesting sites for bird species like Goliath heron and Reef heron [18]. There has been an observed degradation of mangrove trees in the Red Sea region. Some of the major drivers of degradation are the clearing of large areas of mangrove for hardwood, shrimp farming, over grazing by camels, construction activities that alter the water level, and tidal flow. For instance, building dams has significantly lowered fresh water flow into the swamps, thereby, leading to a substantial rise in the level of salinity [19]. In addition to that, mangrove species of this region grow in a particularly harsh environment including: salinity levels of more than 40 ppt, sea surface temperature of more than 31 °C in summers, and no permanent source of freshwater recharge [20].

In the context of the Rabigh lagoon, in 1987, a new road was constructed in the northwest of the study area crossing the mouth of the bay. This road obstructed the recurrence of water flow within the swamp. As a result, the amount of freshwater flow decreased and salinity of the bay area gradually elevated, thereby reducing the productivity of the mangrove trees. By the end of 2012, the road was flooded by sea water, completely opening the bay, and consequently improving the health of mangroves due to the enhancement of water movement.

In order to conserve ecologically sensitive regions, it is inevitable to first monitor their status. Remotely sensed data, especially multi-spectral Landsat data, have been widely used for such purposes [21–25]. Essentially the spectral information from satellite imagery is used to correlate it with the biophysical properties of vegetation cover. Several studies have indicated that the NDVI (Normalized Difference Vegetation Index) of mangrove represents high correlation with biomass and leaf area index [26–30]. Other indices such as Enhanced Vegetation Index (EVI) and Soil Adjusted Vegetation Index (SAVI) are also used to monitor plant health as they are most robust than NDVI. For instance, EVI is sensitive to background canopy variations and does not saturate over high biomass. SAVI accounts for background soil component that can interfere with the vegetation signal [31]. Normalized Difference Moisture Index (NDMI) has been used to detect forest disturbance and recovery since it detects variation in the moisture content of the vegetation. It can differentiate between the moisture content of soil and vegetation and is less sensitive to atmospheric scattering; SWIR (shortwave infrared) used for NDMI can penetrate thin clouds as well [32].

Alamahasheer et al. [33] studied changes in Red Sea mangroves using multi-temporal Landsat imagery, Elsebaie et al. [34] used integrated remote sensing and GIS (geographic information system) approach to identify suitable plantation sites for mangroves on the southern coast of the Red Sea, Kumar et al. [35] surveyed distribution of mangroves along the coast of theRed Sea using NDVI, and Monsef and Smith [36] estimated mangrove cover in the Red Sea using Landsat spectral indices. This study aims to monitor the status of mangrove trees in Rabigh Lagoon by comparing spatial and temporal patterns during the time of obstructed water circulation and after the water circulation was resumed. To our knowledge, this is the first study to use hyper temporal data and trend estimation, over this region, in the form of annual images for the past 34 years.

2. Materials and Methods

2.1. Study Area

Rabigh Lagoon is located on the northeast coast of Red Sea. It, also called as Sharm El Kharar, extends between longitudes 38° E and 39° E and latitudes 22° N and 23° N. Its origin dates to late Pleistocene because of erosion, followed by flooding due to postglacial sea level rise in early Holocene transgression [37–39]. The study region is only 16 km from Rabigh city and occupies an area of about 75 km^2 (Figure 1) with many patches of mangrove communities spreading over an area of 136.7 ha [35]. Surface elevation of the area differs widely from −10 to 58.5 m above sea level. Sediment texture is mostly mud, gravelly sand, and sandy mud. The prevailing climate is dry and tropical. In winters, there is also influx of freshwater from wadis: Rabigh, Rehab, Murayykh, and Al-Khariq.

Tidal range is 20 to 30 cm. Water exchange between the lagoon and the sea takes place mainly via the following process: Water, having temperature of 25 to 30 °C and salinity of 39‰, enters the lagoon as surface inflow. After circulating the lagoon, it exits in the form of subsurface flow with relatively similar temperature and a salinity of 39.8‰ and 40.5‰ during winter and summer, respectively [40]. On average, it takes 15 days for the water to circulate the lagoon, with the speed of flow of 50 cm per second at the entrance 5–20 cm per second inside the lagoon. *Avicennia marina* is the only mangrove species that is found in the lagoon [41]. They mostly have a stunted growth reaching to a height of 2 m. The Rabigh lagoon has suffered from climatic changes and global warming in the form of increased water temperature and sea level rise [42].

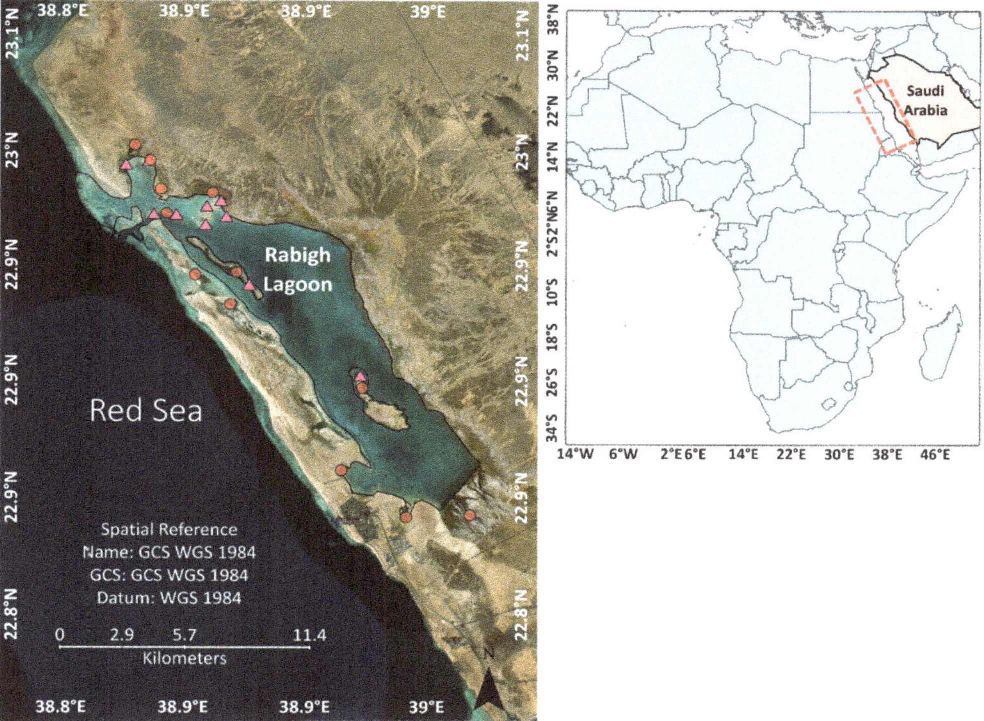

Figure 1. Map of study site, Rabigh Lagoon, located along the coast of Red Sea, Saudi Arabia. Pink triangles show locations for field visits. Red dots show locations from where pixel wise time series were extracted. GCS stands for Geographic Coordinate System. WGS stands for World Geodetic System.

2.2. Trend Analysis Using Time Series

Time series are data points distributed across equal intervals. Time series of vegetation indices have been used for ecological studies such as monitoring forest and land degradation. This study used 30 m annual images, from 1986 to 2019, generated from sensors onboard Landsat 5 (1986 to 2002), Landsat 7 (2003 to 2012), and Landsat 8 (2013 to 2019) to derive the long-term trend component over mangrove forests in the study region. The different indices used were as follows (Table 1): NDVI, EVI, MSAVI, and NDMI. Images for all years belonged to the summer season, which is June to September, depending on data availability and minimum cloud cover. Data were obtained from the online archive of USGS EROS (United States Geological Survey Earth Resources Observation and Science) Center Science Processing Architecture. The final product (normalized indices) downloaded, with path row 170/044 on WRS-2 (World Reference system), was derived using surface reflectance (tier 1) products from Landsat 5 TM (Thematic Mapper), Landsat 7 ETM (Enhanced Thematic Mapper), and Landsat 8 OLI (Operational Land Imager). Surface reflectance products are radiometrically and atmospherically corrected and come along with cloud, shadow, water, and snow masks developed using the CFMASK (C Function of Mask) algorithm. More information about Landsat surface reflectance product characteristics can be found in Schmidt et al. [43] and Vermote et al. [44] findings. All images were rescaled to fall in the range between −1 to 1 by multiplying the values with 0.0001. Images from Landsat 7 between the years 2003 and 2012 with SLC (Scan Line Corrector) failure were gap filled using linear interpolation. All non-vegetated areas and water bodies were masked out from further analysis using a threshold median EVI of 0.08. This threshold was determined by comparing EVI values of vegetation with non-vegetated surface around Rabigh Lagoon. Figure 2 shows spatial distribution of prominent vegetation communities (dark green tones).

Table 1. Spectral indices used for time series analysis.

Spectral Index	Formula	Wavelength (μm)		Reference
NDVI	(NIR − Red)/(NIR + Red)	Red	0.63–0.69	[45]
EVI	G * (NIR − Red)/(NIR + C1 * Red − C2 * Blue + L)	Blue	0.45–0.52	[46]
MSAVI	$(2 * NIR + 1 - sqrt((2 * NIR + 1)^2 - 8 * (NIR - Red))/2$	NIR (near infrared)	0.77–0.9	[47]
NDMI	(NIR − SWIR1)/(NIR + SWIR1)	SWIR1 (shortwave infrared)	1.55–1.75	[31]

NDVI, Normalized Difference Vegetative Index; EVI, Enhanced Vegetative Index; MSAVI, Modified, Soil-Adjusted Vegetative Index; NDMI, Normalized Difference Moisture Index. G = 2, C1 = 6, C2 = 7.5, L = 1.

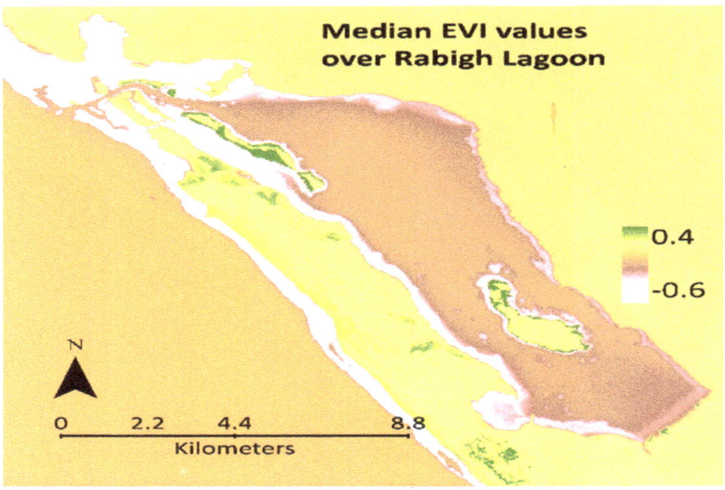

Figure 2. Median EVI over Rabigh Lagoon derived by calculating the median value using all the images between 1986 and 2019.

Workflow in Figure 3 shows how the images were processed and analyzed. Trend estimation was carried out on four (EVI, MSAVI, NDVI, and NDMI) annual raster time series covering the period of the last 34 years. Temporal changes in indices were interpreted as follows: increasing EVI, MSAVI, and NDVI meant the canopy is more productive (green) and in a healthy state. This is due to higher reflection of NIR (near infrared) by the spongy mesophyll of the all the leaves in the canopy and absorption of red wavelength by chlorophyll. In the case of NDMI, the SWIR1 reflection tells about the vegetation water content and spongy mesophyll structure. Higher water content would mean more absorption of SWIR1, thus leading to higher NDMI [48]. Slope of the yearly trend was calculated by linear least square regression where the index values act as response variable against the predictor variable, which in this case is time. Statistical significance of the regression coefficient was determined through Mann-Kendall trend test. The analysis was done using the "greenbrown" library [49] in RStudio (version 1.2.5033, Affero General Public License v3, Boston, MA, USA). The trend calculation algorithm in the "greenbrown" library offers the possibility to compute breaks in the time series by controlling the parameter h. Statistically, a breakpoint in the time series occurs when there is a structural change in the regression parameters before and after the break [50]. Ordinary Least Square Moving Sum (OLS-MOSUM) test is applied to detect whether any breaks exist in the time series. The detected break in a signal might depict an abrupt change such as clearcutting, logging, forest fire, landslide, and land management practices that might cause the index values to change with a large magnitude. However, breaks can also be induced by data artifacts [51]. For this study, the parameter h, which is the minimum time between detected breaks, was set to 5 years (0.15) for a 34-year time span. This means only the most significant abrupt change, if it exists, in the five-year period will be recorded. The regression analysis resulted in four trend maps showing areas with monotonic upward (greening) and downward (browning) patterns.

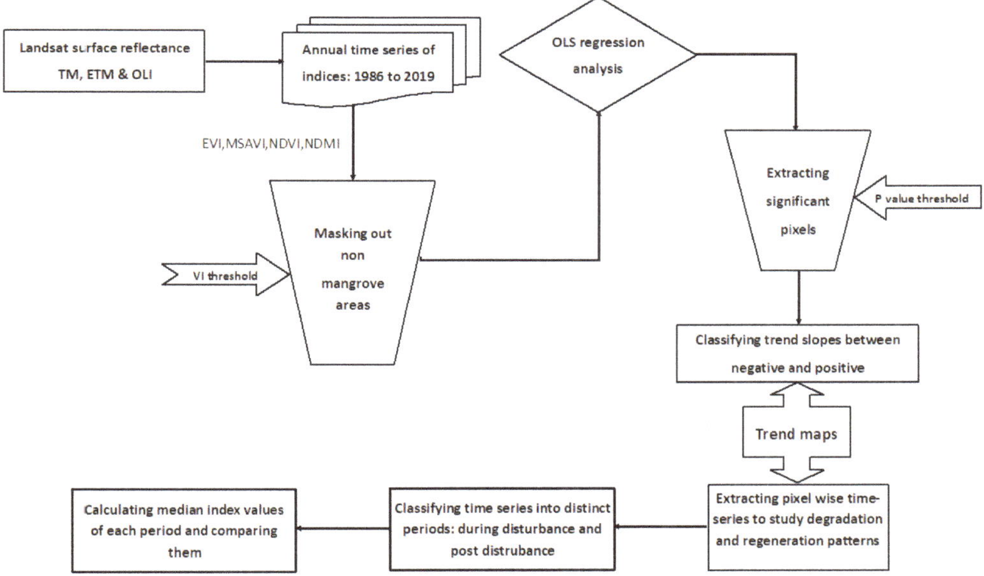

Figure 3. Framework for data acquisition and analysis.

All the spatial mapping was carried in ArcGIS Pro (version 2.6.3, Esri, Redlands, CA, USA). Using the trend maps, pixel-wise time series were extracted for each index and were analyzed as follows: to depict the health of mangroves, each time series was divided into two periods; during road obstruction (disturbed period) and after demolition of the road (recovery period). The disturbed period was represented by the years 1987 to 2012, whereas recovery period was represented by years 2013 to 2019. The median was calculated for each period per site and subsequently compared. If the difference between the median index values between the two periods was positive that was interpreted as recovery. Additionally, for easy comparison, the differences between disturbed and recovery periods for all twelve sites were further averaged for each index. Similar calculations were made for the median of raster images. Kruskal-Wallis test was also performed in RStudio to identify any significant differences between the medians of both periods using median values for each site per index. Kruskal-Wallis is a non-parametric approach to detect any significant differences in a continuous dependent variable (median index values) against a categorical independent variable (period of disturbance and post disturbance recovery).

3. Results

Trend estimations from all four indices were analyzed for their spatial explicitness and temporal development. Figure 4 shows the spatial trend patterns computed by the regression algorithm. Only significant trends with p-values less than and equal to 0.05 (95% significance level) are shown. All trend slopes lie within the range of -0.01 to 0.01. Areas on trend maps (Figure 4) with bright orange/yellow to green colors indicate the range of weak positive (<0.01) to strong positive (0.01) trend, while areas with dark orange to red indicate the range for weak negative (<-0.01) to strong negative trends (-0.01). According to the histograms in Figure 5a–d, most pixels had a weak positive trend for all indices falling in the range of 0.001 to 0.005. MSAVI showed higher variation in the calculation of trend as can be seen by a greater number of bars. In addition to that, the number of pixels with significant positive trends (more than 0.001) was the highest for MSAVI (Figure 5b). This presents a good case for using MSAVI over regions with sparse vegetation since it can prevent the influence of background soil. The number of pixels with negative trends (slope value more than -0.001) was highest for NDMI, indicating water-stressed mangroves (Figure 5d). The increasing trend shown in Figure 4 was mostly observed over natural areas with mangrove forests depicting subtle changes. The vegetation around Rabigh lagoon is dominated by mangrove trees [41]. Just near the coast, on the south west of the lagoon, there is a cluster of pixels with a strong positive trend over a region that has undergone modification and land use in terms of managed tree plantation on an area that was previously bare soil. Stronger greening patterns were observed for mangroves inside the lagoon compared to those for the mangroves just at the north western border of the lagoon. Of all four indices, the NDVI was the least reliable. Although overall patterns were similar, the NDVI does not perform well in sparsely vegetated areas, especially where highly heterogeneous pixels are influenced by surrounding bright soil.

Different plots (Figure 6) of calculated trend slopes for all four indices show correlation between EVI, MSAVI, NDVI, and NDMI and their statistical distribution. Overall correlation values between EVI, MSAVI, and NDVI are quite high, up to 0.98. On the other hand, lower values of correlation between NDMI with other indices were observed, with the maximum reaching up to 0.58. Nonetheless, all indices showed a dominant greening pattern. The box plots and QQ (quantile-quantile) plots show that all indices follow similar statistical distribution. The prominent presence of extreme values (especially positive) is indicated by data points falling on the line in the middle and deviating away from the line (above blue line for positive values and below blue line for negative values) forming heavy tails in the QQ plots. This is also reflected by the data points falling outside the upper and lower whisker in the box plot.

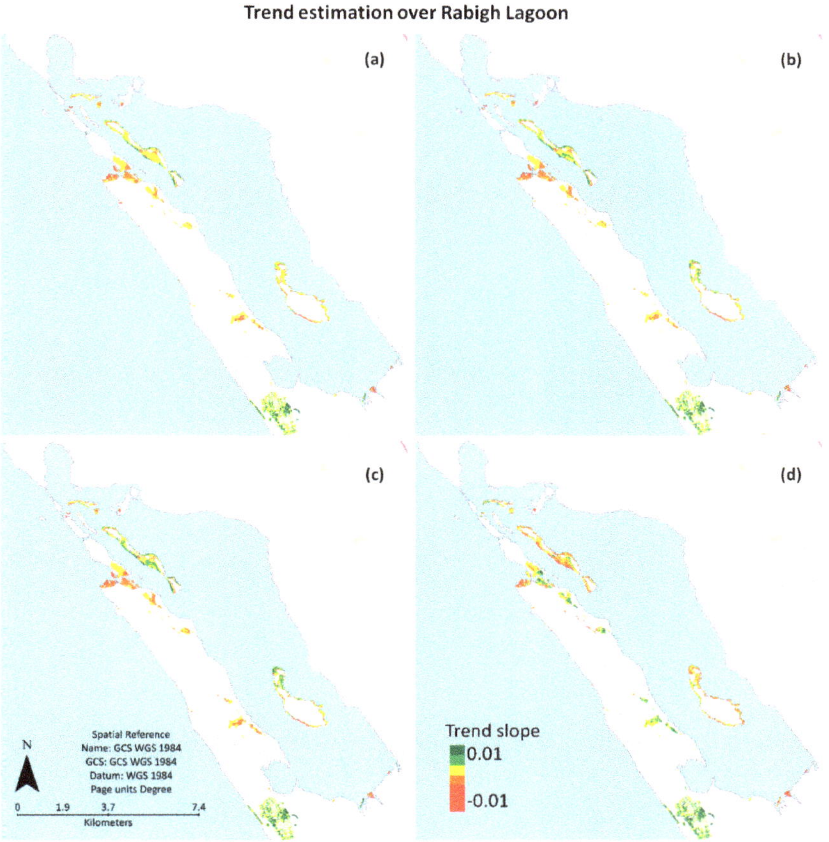

Figure 4. Regression slopes derived from the trend algorithm for (**a**) EVI, (**b**) MSAVI, (**c**) NDVI, and (**d**) NDMI.

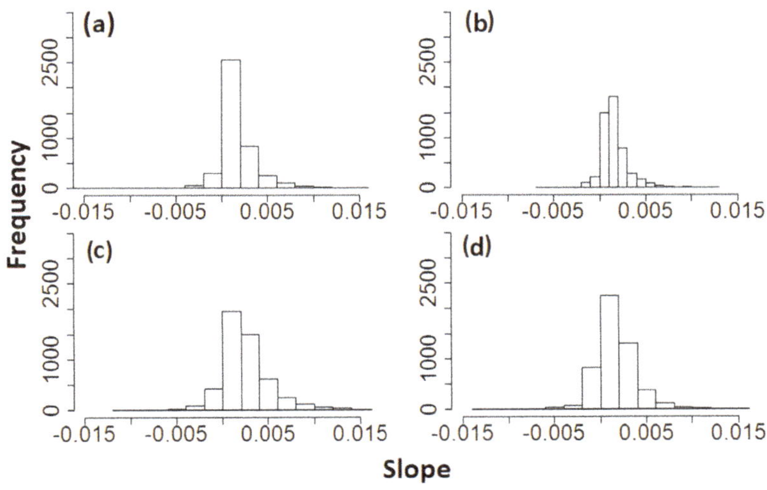

Figure 5. Histograms for trend slopes derived from (**a**) EVI, (**b**) MSAVI, (**c**) NDVI, and (**d**) NDMI.

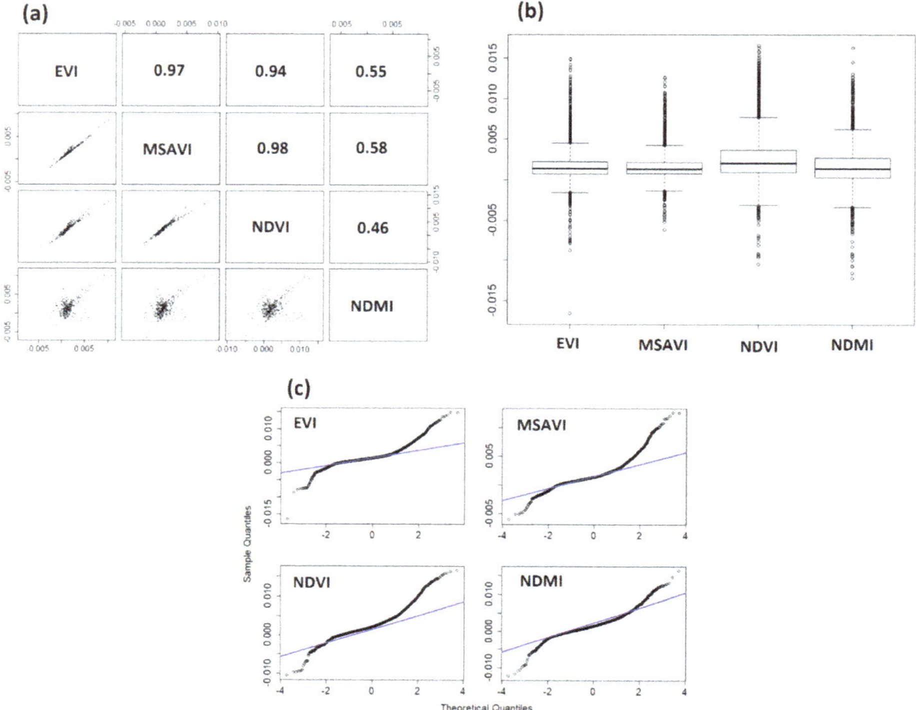

Figure 6. Pair wise scatter plots (**a**), boxplot (**b**), and QQ (quantile-quantile) plots (**c**) showing the correlation and statistical distribution of trend slopes derived from EVI, MSAVI, NDVI, and NDMI.

Figure 7 shows the extracted time series over various sites with mangrove trees. A total of 12 time series were extracted. All the graphs show a very subtle development of index values. NDMI shows lower values than the other indices, which reflects water stress for plants largely triggered by obstructed water flow. For all indices, the end of time series can be seen showing higher values, and they stay above the values of those in the beginning of the time series (discussed further in Figure 8). That is why there is an observed monotonic upward trend. Low index values in 2005 were accounted for by the poor quality according to the "pixel quality assurance" band. Therefore, it was considered as noise. Outliers were considered as noise in this study. As long as noise did not result in break detection, cause regression parameters to be unstable, or lead to trend shifts, it was ignored from further investigation. In addition, any rise and fall in the signal between consecutive years was not interpreted as a trend because it was assumed that the mangrove ecosystem takes time to respond to external factors. Moreover, the lag period between the cause and effect has to be kept in mind when detecting vegetation response with satellite instruments. Local scale differences between the time series of the VIs (vegetation index) and moisture index can be attributed to the spectral differences where the VI used the IR (infrared) and visible (red and blue) part of the spectrum for highlighting the greenness of the vegetation. On the other hand, NDMI use only the infrared waves (IR and SWIR) characterizing the wetness of the vegetation.

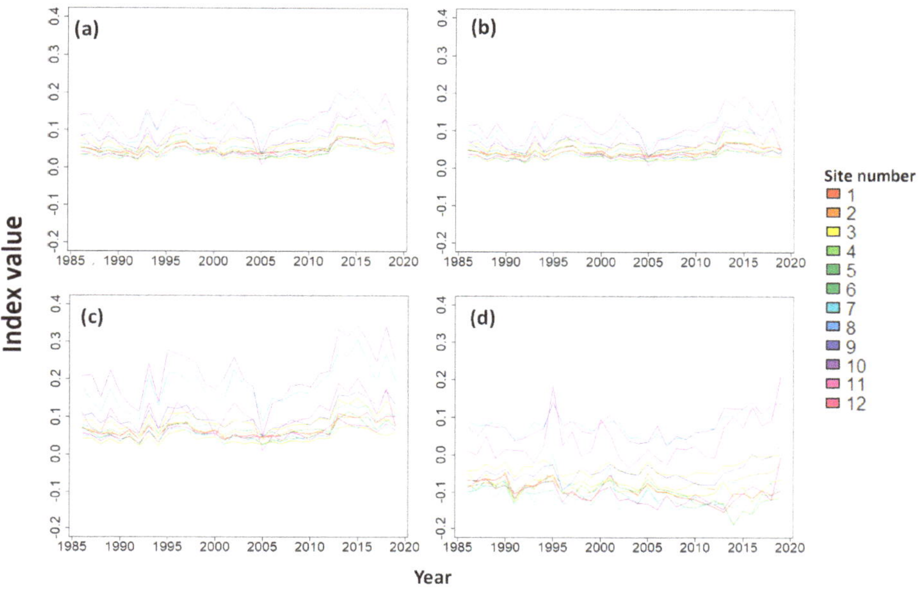

Figure 7. Time series between 1986 and 2019 extracted from different mangrove sites (for all four indices: (**a**) EVI, (**b**) MSAVI, (**c**) NDVI, and (**d**) NDMI.

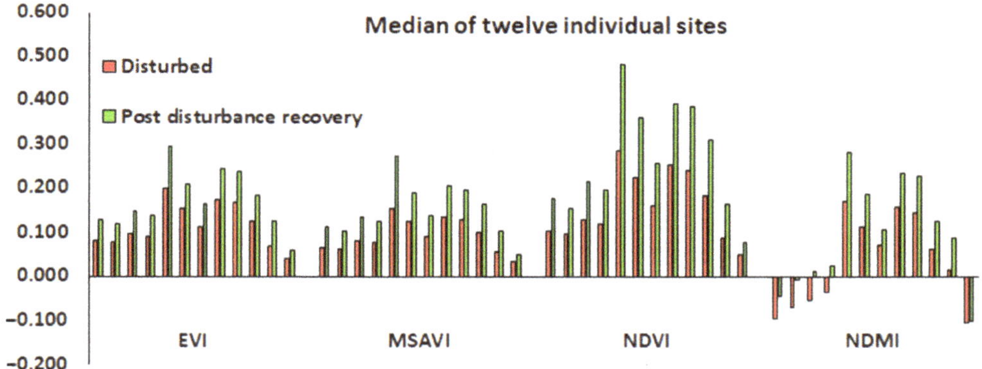

Figure 8. Median index values for disturbed and post disturbance recovery of mangroves for all twelve sites (Disturbance: 1987–2012, Recovery: 2013–2019).

In Figure 8, median values for each selected site can be seen changing during the two classified periods of time series. According to the output of Kruskal-Wallis test with p-value of 0.0006918 (against significance level of 0.05), there are significant differences in the median values of disturbance period from the recovery period. For all sites during the disturbance period (1987–2012), median values are lower, showing mangrove communities in an unhealthy state. Post disturbance (2013–2019), after demolishing the road, a gradual recovery of the mangroves was observed, which was reflected in the increased median index values compared to the values in disturbed state. Figure 9 shows the spatial pattern of median differences for the whole study site, which also came out to be positive. The average differences (Table 2) between the two periods were similar for EVI and MSAVI, whereas for NDVI, it was the highest. The NDVI usually shows higher values because it tends to saturate over high biomass; therefore, the larger difference was not interpreted as a higher recovery. In the case of NDMI, the difference was only slightly larger than the difference of EVI and MSAVI. This could be a random occurrence. However, it is worthy to mention that NDMI is better at identifying forest disturbance and recovery because SWIR is more influenced by canopy moisture content [52]. For some sites (38.8361426° E 22.9962740° N, 38.9113980° E 22.8789328° N, 38.9140997° E 22.8773751° N), NDMI shows negative values even in the recovery period, which could mean higher water stress. In addition, negative NDMI values also mean low canopy cover, which was illustrated by visualizing high-resolution Google Earth imagery.

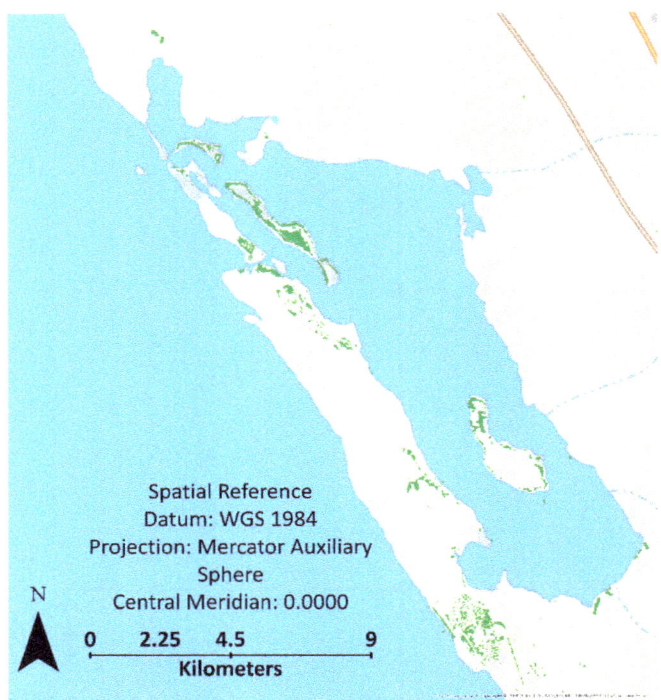

Figure 9. Spatial pattern of post disturbance recovery (show in green).

Table 2. Averaged median values of disturbed and recovery period and their difference.

Index	Disturbance	Recovery	Difference
EVI	0.104	0.155	0.052
MSAVI	0.086	0.136	0.051
NDVI	0.143	0.235	0.092
NDMI	0.038	0.096	0.066

This study relates degradation and recovery patterns in response to obstructed water recharge, resulting in disturbed levels of salinity, dissolved oxygen, and pH. Tidal and water chemistry are two main parameters considered to effect mangrove productivity [53]. Therefore, it is helpful to consider these factors when interpreting trend patterns. Under higher-salinity and low-nutrient conditions, mangroves increase the process of transpiration to excrete the salt, which could lead to less growth or productivity. Figure 10 shows the aforementioned parameters of the lagoon for the year 2014. The data were taken from the study of Youssef and Sorogy [54] and mapped to visualize the spatial pattern. Since no such data are available during the disturbance period, a direct comparison cannot be made. Nevertheless, values in Figure 10 do suggest that mangroves are returning to a healthy state because the parameters in the surrounding lagoon water are within the normal range (mangroves in Rabigh are acclimatized to 40 ppt salinity) [20].

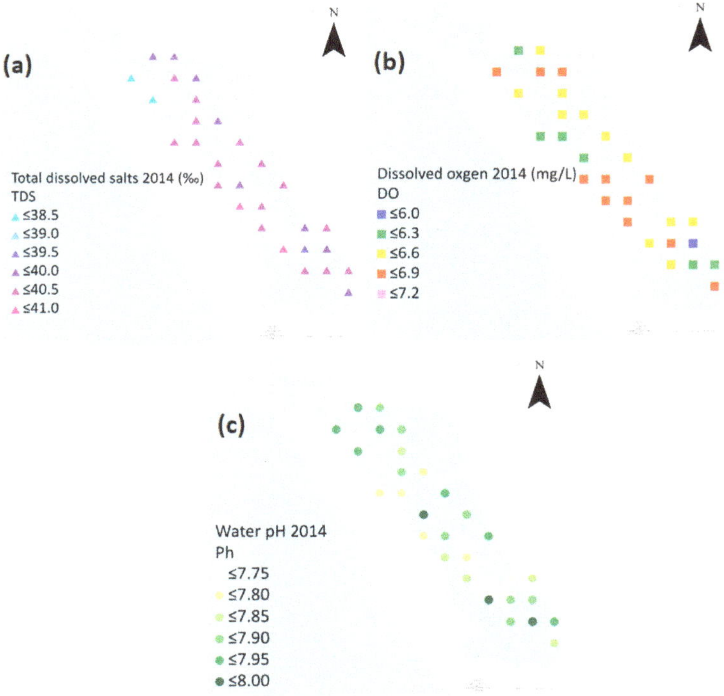

Figure 10. Surface water parameters (**a**) salinity or dissolved salts, (**b**) dissolved oxygen and (**c**) pH. The sampling and estimation was done in 2014 [55].

Rainfall and temperature data, provided by CRUTS (Climate Research Unit gridded Time Series), were also used to assess their impact on degradation and regeneration patterns. The data were produced through the interpolation of observations from weather stations. Monthly data was downloaded, and the trend was estimated through linear least square regression on annually aggregated series against a 95% significance level. According to Figure 11, an increasing trend in temperature and non-significant trend in precipitation was observed. Whereas it could be contemplated that a combination of both climatic parameters could have had an influence on mangroves, this cause-and-effect relation must be proceeded on with extreme caution. Firstly, the data had a very coarse spatial resolution. For pixel-to-pixel correlation, climatic and index data must be of the same spatial resolution. Rainfall and temperature were recorded from the monitoring station in Jeddah that is at a distance of 141 km from Rabigh. While such datasets might be useful to monitor the impacts of major climatic events and general trend patterns in the context of globally rising temperatures, it might not always be helpful in reflecting a spatially and temporally explicit situation. This is especially the case where a combination (multivariate) of factors have a nonlinear effect and linear statistical correlations might not be enough.

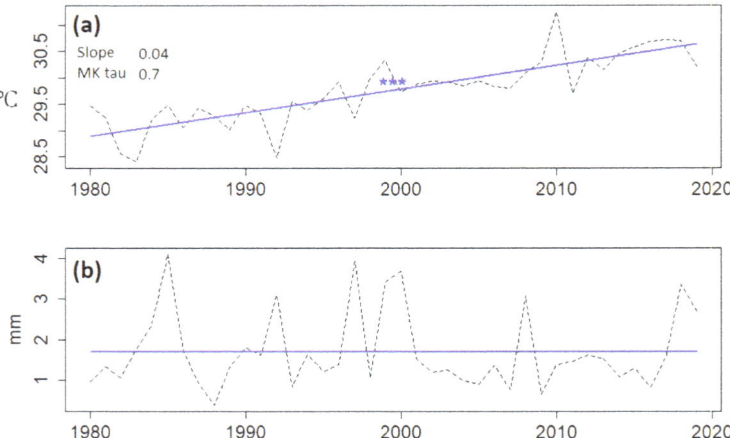

Figure 11. Trend (blue line) analysis for annual average (**a**) temperature and (**b**) rainfall. Trend for rainfall came out to be non-significant, therefore slope and MK (Mann-Kendall) tau values are not shown. The symbol * shows the trend is significant, with three * meaning highly significant having *p* value of less than 0.000001.

4. Discussion

The deterioration of mangrove species has been observed at a regional and global scale. According to Al Shiekh [55], there has been a loss of 46% of vegetation cover between 1987 and 2002 in the Red Sea coast at the Jazan region, Saudi Arabia. The reasons discussed for this loss are demographic growth and aquaculture. On the contrary, a recent study by Almahasheer et al. [33] reported an expansion of 12% in mangrove area along the Red Sea. This expansion was observed over 41 years from 1972 to 2013 based on multi-temporal Landsat data analysis. Although our study investigated only a small part of the Red Sea, it does correlate with the findings of Almahasheer et al. [33] such that our study showed a monotonic upward trend for mangroves in the form of regeneration after the road in the north west of the lagoon was dismantled. Our study used the approach of "intercomparison" for indirect validation of the observed trends. Intercomparison involves [56] using multiple data products to draw out simplified estimations of major spatial and temporal patterns. The multiple data products in this study were the four indices using different band combination to assess the temporal change in the spectral response.

Mangroves show a lot of tolerance to various stress factors they are exposed to [57]. The generated time series (Figure 7) of annual records of EVI, MSAVI, NDVI, and NDMI over twelve sites of mangrove habitats in the Rabigh lagoon from 1986 to 2019 showed an interesting pattern of ecological dynamics of mangrove population. There are very subtle changes taking place in the region, where at some sites, mangroves tend to be stable through a loss and gain strategy, and at others, they show a declining tendency. According to Vovides et al. [58], mangrove regeneration post disturbance in arid and semi-arid regions could prove to be a challenge. This is especially the case where plantation activities alone, without eliminating the stress factor, will not bring desired results. Twilley et al. [59] investigated similar hydrological stress on mangroves in CGSM lagoon (Cienaga Grande de Santa Marta) Colombia, brought on by coastal highway construction. The authors highlighted that the recovery rate of the mangrove wetlands depends on the intensity and persistence of the disturbance, type and age of species, and natural tree mortality, where adult trees after dying leave gaps in the canopy giving space for new seedlings to grow. Based on the study of Twilley et al. [59], natural hydrological restoration and mangrove plantations were the main assessed factors important for mangrove recovery.

In the region of Rabigh, the prominent limiting factors for mangrove growth are optimum hydrological conditions in the form of mainly water circulation from the open sea and limited rainfall. A sand pathway, hindering water circulation, has been present between 1987 and 2012. Since mangrove forests are wetlands and have a deep root system, persistent water obstruction may take a few years to have a prolonged and visible (such as stunted growth and yellowing of leaves) and detectable impact on mangrove's health [60]. This is so because mangrove roots are efficient in absorbing water from sediments. Moreover, since Rabigh Lagoon is surrounded by inland desert, mangrove species there are adapted to dry and warm conditions and will show resilience to drought-like conditions. It should be kept in mind though that at the physiological level, degradation might be immediately detected. The lag discussed in this paper is solely in terms of the satellite detection of vegetation status when viewing the canopy greenness and wetness. The sand passageway was partly flooded by water in 1990, after which the lagoon was connected to the sea through manually laid downpipes (Figure 12). With the passage of time, these pipes were blocked by sedimentation. As a result, there was a slow and reduced water re-charge into the lagoon. During this period, the mangroves were gradually recovering while the degradation processes were still active.

Figure 12. Pipes laid down under the sand passage to allow water circulation.

In 2012, the sand passageway was completely demolished, and the lagoon was re-connected with the open sea. As a result, there was unobstructed natural water circulation and also an increased nutrient supply. This is reflected by increasing index values for all indices in all twelve sites, especially from 2013 onward where the values are the highest during the whole evaluation period. Three sites in particular were visited as part of an ongoing project in Rabigh Lagoon since the 1990s and therefore are presented as case studies in the following sections. Figures 13–15 show results from pixel-based trend analysis for EVI, MSAVI, NDVI, and NDMI for the three locations. They are further discussed below. All three sites present a case study of disturbance (due to road construction) and recovery after the road was demolished and the effect of various stress factors was diminished.

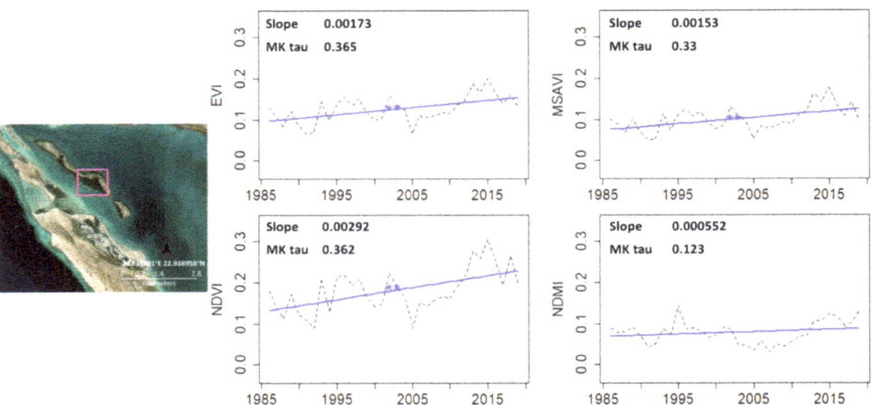

Figure 13. Trend (blue line) analysis for site in the north (enclosed in pink box). The symbol * shows the trend is significant, with two * meaning *p* value of equal or less than 0.001.

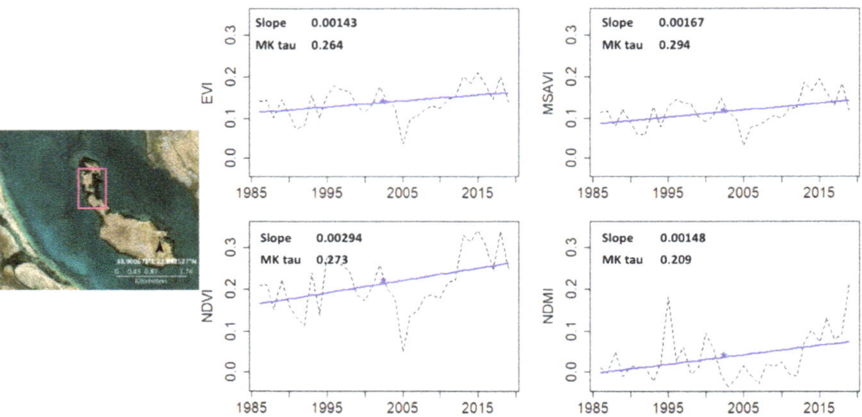

Figure 14. Trend (blue line) analysis for site in the south (enclosed in pink box). The symbol * shows the trend is significant, with one * meaning *p* value of equal or less than 0.05.

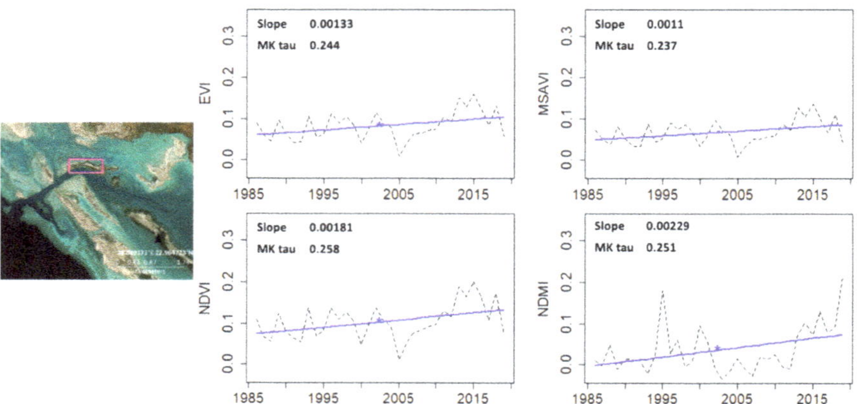

Figure 15. Trend (blue line) analysis for site in extreme northwest (enclosed in pink box). The symbol * shows the trend is significant, with one * meaning *p* value of equal or less than 0.05.

4.1. North of Lagoon

This site, as shown in Figure 13, is in the northern part of the lagoon facing a skewed, deep entrance to the open sea. Due to this, only the tiny deep channel allows water circulation between the lagoon and the open sea. The habitat at this location is mixed between coral reefs and mangroves. Mangroves are scattered in patches. This area has suffered from extreme wind and tides that led to massive erosion; therefore, mangrove trees were under high environmental stress. Slope values for EVI, MSAVI, and NDVI range from 0.0015 to 0.003. The MK (Mann-Kendall) tau correlation coefficient came out to be the highest for EVI, MSAVI, and NDVI, which meant that around 35% variability in VI is explained by time. However, in the case of NDMI, low tau and slope values indicate a weak relation between trend and time and weak trend, respectively. This could indicate that the moisture content of mangroves at this particular location might not be a driving factor for degradation and recovery patterns as much as chlorophyll content is. No breaks were detected, which means there has been a gradual upward trend.

4.2. South of Lagoon

This site is present in the second island toward the southern part of the lagoon where freshwater flow is mainly supplied by rain and flood. According to Figure 14, the slope values lie between 0.0015 and 0.003. Only gradual change was observed. MK tau values showed that 20% to 30% variation in index values was explained by time. This site was under sedimentation process, receiving more mud and clay from runoff. The increase of sedimentary conditions may have affected water circulation and increased the deposition of organic matter causing lower index values during the disturbance period.

4.3. Extreme Northwest of Lagoon

Located in the Camel Island facing the third entrance of water, it is relatively a small channel. It can be seen from Figure 15 that no abrupt changes were detected by the regression algorithm. At low tide, camels used to cross the shallow water toward the island for grazing. This site has suffered the most degradation, as the index values are lower compared to the other two sites. The calculated trend for EVI, MSAVI, and NDVI was in the range of 0.001. On the contrary, slope value (0.002) and tau value (25%) were the highest for NDMI for this site, suggesting that moisture was the dominant limiting factor. Figure 16 shows photos taken from the field visit conducted at this particular location during various stages of an ongoing project. The temporal variability of mangrove vegetation can be visualized; there is evident degradation in the 1990s followed by a return to a healthy state after natural water flow was resumed.

Figure 16. Photos from field visits: (**a**) 1993 showing disturbed mangroves and (**b**) 2019 showing recovering mangroves at site in extreme northwest.

For all three sites, no breaks were found as result of abrupt changes in the index values. This indicates there is no clearcutting and large-scale urbanization that would cause a sudden drop in the index value with at least a change in magnitude exceeding 0.2 units. Mostly the changes are subtle and persistent. That is why despite the curve having a period of low values, the slope line shows up as a long-term pattern of increasing trend. This study only investigated EVI, MSAVI, NDVI, and NDMI as an indicator for mangrove disturbance and recovery. Other parameters such as water salinity gradients, dissolved oxygen, and other water quality parameters could not be analyzed with the same temporal frequency as the indices. A change in population or community structure occurring at the boundaries of habitats is also an essential aspect that is difficult to monitor with medium-resolution imagery. Additionally noteworthy here is the influence of marine erosion on the stability of mangroves, especially for smaller patches of mangroves, as is the case for Rabigh. It is suggested to consider these additional drivers in future studies regarding mangroves of the Red Sea and particularly in the area of Rabigh lagoon. In addition, to better understand vegetation recovery, experimental setups or projects involving mangrove restoration could be devised to study and prioritize drivers essential for regeneration processes.

5. Conclusions

This study attempted to monitor mangrove disturbance and the post-disturbance recovery pattern in Rabigh lagoon along the coast of the Red Sea through trend analysis of various indices. Moreover, observed patterns were analyzed in the context of road construction that hindered water circulation through the lagoon. Our aim was to capture long-lived processes as opposed to temporary and short-lived changes. The outcome of this study showed that time series analysis of selected vegetation and moisture indices (EVI, MSAVI, NDVI and NDMI) was able to capture a pattern of disturbance and recovery. Even if the changes were gradual, they could still be identified. It is suggested to use MSAVI and NDMI to assess ecological disturbance and recovery, particularly for sparsely vegetated and arid areas. However, validating observed patterns using only the regression slope proved to be a challenge. This is so because the response of a natural ecosystem to gradual changes is complex. Integrating rain and temperature data in our interpretation scheme was only partially helpful, as it aided in explaining general patterns, but climatic anomalies were not able to account for extreme values in the index time series. Future studies must integrate the estimation of surface water parameters at constant time intervals to relate various stress factors with mangrove response. Furthermore, extensive and regular field visits to visualize the condition of mangrove stands could also help validate the observed patterns.

Author Contributions: Conceptualization, M.O.A. and W.R.K.; methodology, W.R.K. and S.M.; formal analysis, S.M.; data, S.M.; writing—original draft preparation, M.O.A. and W.R.K.; writing—review and editing, S.M.; visualization, M.O.A.; project administration, M.O.A.; funding acquisition, M.O.A. All authors have read and agreed to the published version of the manuscript.

Funding: This project was funded by the National Plan for Science, Technology, and Innovation (MAARIFAH), King Abdulaziz City for Science and Technology, the Kingdom of Saudi Arabia (award number 14-ENV263-03).

Acknowledgments: This project was funded by the National Plan for Science, Technology, and Innovation (MAARIFAH), King Abdulaziz City for Science and Technology, the Kingdom of Saudi Arabia (award number 14-ENV263-03). The authors also acknowledge with thanks the Science and Technology Unit, King Abdulaziz University for technical support.

Conflicts of Interest: The authors declare no conflict of interest.

References

1. Guangyi, Z. Influences of tropical forest changes on environmental quality in Hainan province, PR of China. *Ecol. Eng.* **1995**, *4*, 223–229. [CrossRef]
2. Costanza, R.; d'Arge, R.; De Groot, R.; Farber, S.; Grasso, M.; Hannon, B.; Limburg, K.; Naeem, S.; O'neill, R.V.; Paruelo, J. The value of the world's ecosystem services and natural capital. *Nature* **1997**, *387*, 253–260. [CrossRef]
3. Khan, W.R.; Nazre, M.; Zulkifli, S.Z.; Kudus, K.A.; Zimmer, M.; Roslan, M.K.; Mukhtar, A.; Mostapa, R.; Gandaseca, S. Reflection of stable isotopes and selected elements with the inundation gradient at the Matang Mangrove Forest Reserve (MMFR), Malaysia. *Int. For. Rev.* **2017**, *19*, 1–10. [CrossRef]
4. Lee, H.-Y.; Shih, S.-S. Impacts of vegetation changes on the hydraulic and sediment transport characteristics in Guandu mangrove wetland. *Ecol. Eng.* **2004**, *23*, 85–94. [CrossRef]
5. Lewis, R.R., III. Ecological engineering for successful management and restoration of mangrove forests. *Ecol. Eng.* **2005**, *24*, 403–418. [CrossRef]
6. Willis, J.M.; Hester, M.W.; Shaffer, G.P. A mesocosm evaluation of processed drill cuttings for wetland restoration. *Ecol. Eng.* **2005**, *25*, 41–50. [CrossRef]
7. Sánchez-Carrillo, S.; Sánchez-Andrés, R.; Alatorre, L.C.; Angeler, D.G.; Álvarez-Cobelas, M.; Arreola-Lizárraga, J.A. Nutrient fluxes in a semi-arid microtidal mangrove wetland in the Gulf of California. *Estuar. Coast. Shelf Sci.* **2009**, *82*, 654–662. [CrossRef]
8. Fickert, T. To Plant or Not to Plant, That Is the Question: Reforestation vs. Natural Regeneration of Hurricane-Disturbed Mangrove Forests in Guanaja (Honduras). *Forests* **2020**, *11*, 1068. [CrossRef]
9. Khan, W.R.; Rasheed, F.; Zulkifli, S.Z.; Kasim, M.R.B.M.; Zimmer, M.; Pazi, A.M.; Kamrudin, N.A.; Zafar, Z.; Faridah-Hanum, I.; Nazre, M. Phytoextraction Potential of Rhizophora Apiculata: A Case Study in Matang Mangrove Forest Reserve, Malaysia. *Trop. Conserv. Sci.* **2020**, *13*, 1940082920947344. [CrossRef]
10. Kathiresan, K.; Bingham, B.L. Biology of mangroves and mangrove ecosystems. *Adv. Mar. Biol.* **2001**, *40*, 84–254.
11. Spalding, M.; Blasco, F.; Field, C. *World Mangrove Atlas*; International Society for Mangrove Ecosystems: Okinawa, Japan, 1997. Available online: www.environmentalunit.com/Documentation/04%20Resources%20at%20Risk/World%20mangrove%20atlas.pdf (accessed on 15 October 2020).
12. Dahdouh-Guebas, F.; Mathenge, C.; Kairo, J.G.; Koedam, N. Utilization of mangrove wood products around Mida Creek (Kenya) amongst subsistence and commercial users. *Econ. Bot.* **2000**, *54*, 513–527. [CrossRef]
13. Ruiz-Luna, A.; Berlanga-Robles, C.A. Land use, land cover changes and coastal lagoon surface reduction associated with urban growth in northwest Mexico. *Landsc. Ecol.* **2003**, *18*, 159–171. [CrossRef]
14. Ruiz-Luna, A.; Acosta-Velázquez, J.; Berlanga-Robles, C.A. On the reliability of the data of the extent of mangroves: A case study in Mexico. *Ocean Coast. Manag.* **2008**, *51*, 342–351. [CrossRef]
15. Saenger, P.; Khalil, A.S.M. Regional Action Plan for the Conservation of Mangroves. In *PERGA/GEF Regional Action Plan for the Conservation of Marine Turtles, Seabirds and Mangroves in the Red Sea and Gulf of Aden*; Technical Series 12; PERSGA: Jeddah, Saudi Arabia, 2007; pp. 101–129.
16. Khalil, A.S.M.; Krupp, F. Fishes of the mangrove ecosystem. In *Comparative Ecological Analysis of Biota and Habitats in Littoral and Shallow Sublittoral Waters of the Sudanese Red Sea*; Report for the Period of Apr 1991–Dec 1993; Faculty of Marine Science and Fisheries, Port Sudan, and ForschungsinstitutSenckenberg: Frankfurt, Germany, 1994.
17. Wilkie, M.L. *Mangrove Conservation and Management in the Sudan*; Consultancy Report. Based on the Work of Ministry of Environment and Tourism. FOL/CF Sudan. FP:GCP/SUD/O47/NET; FAO: Khartoum, Sudan, 1995; 92p.
18. Ormond, R.F.G. Management and conservation of Red Sea habitats. In *The coastal and marine environments of the Red Sea, Gulf of Aden and Tropical Western Indian Ocean. Proceedings of Symposium, Khartoum (January). The Red Sea and the Gulf of Aden Environmental Programme Jeddah*; 1980; Volume 2, pp. 135–162. Available online: http://www.reefbase.org/resource_center/publication/pub_4287.aspx (accessed on 15 October 2020).
19. Mandura, A.S. A mangrove stand under sewage pollution stress: Red Sea. *Mangroves Salt Marshes* **1997**, *1*, 255–262. [CrossRef]
20. Raitsos, D.E.; Hoteit, I.; Prihartato, P.K.; Chronis, T.; Triantafyllou, G.; Abualnaja, Y. Abrupt warming of the Red Sea. *Geophys. Res. Lett.* **2011**, *38*. [CrossRef]
21. Rasolofoharinoro, M.; Blasco, F.; Bellan, M.F.; Aizpuru, M.; Gauquelin, T.; Denis, J. A remote sensing based methodology for mangrove studies in Madagascar. *Int. J. Remote Sens.* **1998**, *19*, 1873–1886. [CrossRef]
22. Gao, J. A comparative study on spatial and spectral resolutions of satellite data in mapping mangrove forests. *Int. J. Remote Sens.* **1999**, *20*, 2823–2833. [CrossRef]
23. Blasco, F.; Aizpuru, M. Mangroves along the coastal stretch of the Bay of Bengal: Present status. *Ind. J. Mar. Sci.* **2002**, *31*, 9–20.
24. Saito, H.; Bellan, M.F.; Al-Habshi, A.; Aizpuru, M.; Blasco, F. Mangrove research and coastal ecosystem studies with SPOT-4 HRVIR and TERRA ASTER in the Arabian Gulf. *Int. J. Remote Sens.* **2003**, *24*, 4073–4092. [CrossRef]
25. Alatorre, L.C.; Sánchez-Andrés, R.; Cirujano, S.; Beguería, S.; Sánchez-Carrillo, S. Identification of mangrove areas by remote sensing: The ROC curve technique applied to the northwestern Mexico coastal zone using Landsat imagery. *Remote Sens.* **2011**, *3*, 1568–1583. [CrossRef]
26. Jensen, J.R.; Lin, H.; Yang, X.; Ramsey, E., III; Davis, B.A.; Thoemke, C.W. The measurement of mangrove characteristics in southwest Florida using SPOT multispectral data. *Geocarto Int.* **1991**, *6*, 13–21. [CrossRef]

27. Ramsey, E.W., III; Jensen, J.R. Remote sensing of mangrove wetlands: Relating canopy spectra to site-specific data. *Photogramm. Eng. Remote Sens.* **1996**, *62*, 939–948.

28. Green, E.P.; Mumby, P.J.; Edwards, A.J.; Clark, C.D.; Ellis, A.C. Estimating leaf area index of mangroves from satellite data. *Aquat. Bot.* **1997**, *58*, 11–19. [CrossRef]

29. Green, E.P.; Clark, C.D.; Mumby, P.J.; Edwards, A.J.; Ellis, A.C. Remote sensing techniques for mangrove mapping. *Int. J. Remote Sens.* **1998**, *19*, 935–956. [CrossRef]

30. Green, E.; Mumby, P.; Edwards, A.; Clark, C. *Remote Sensing: Handbook for Tropical Coastal Management*; United Nations Educational, Scientific and Cultural Organization (UNESCO): Paris, France, 2000.

31. Huete, A.; Didan, K.; Miura, T.; Rodriguez, E.P.; Gao, X.; Ferreira, L.G. Overview of the radiometric and biophysical performance of the MODIS vegetation indices. *Remote Sens. Environ.* **2002**, *83*, 195–213. [CrossRef]

32. Gao, B.-C. NDWI—A normalized difference water index for remote sensing of vegetation liquid water from space. *Remote Sens. Environ.* **1996**, *58*, 257–266. [CrossRef]

33. Almahasheer, H.; Aljowair, A.; Duarte, C.M.; Irigoien, X. Decadal stability of Red Sea mangroves. *Estuar. Coast. Shelf Sci.* **2016**, *169*, 164–172. [CrossRef]

34. Elsebaie, I.H.; Aguib, A.S.H.; Al Garni, D. The Role of Remote Sensing and GIS for Locating Suitable Mangrove Plantation Sites Along the Southern Saudi Arabian Red Sea. *Int. J. Geosci.* **2013**, *4*, 471–479. [CrossRef]

35. Kumar, A.; Khan, M.A.; Muqtadir, A. Distribution of mangroves along the Red Sea coast of the Arabian Peninsula: Part-I: The northern coast of western Saudi Arabia. *Earth Sci. India* **2010**, *3*.

36. Abd-El Monsef, H.; Smith, S.E. A new approach for estimating mangrove canopy cover using Landsat 8 imagery. *Comput. Electron. Agric.* **2017**, *135*, 183–194. [CrossRef]

37. Braithwaite, C.J.R. *Geology and Palaeogeography of the Red Sea Region*; Edwards, A.J., Head, S.M., Eds.; Red Sea (Key Environments); Pergamon Press: Oxford, UK, 1987; pp. 22–44.

38. Brown, G.F.; Schmidt, D.L.; Huffman, A.C., Jr. *Geology of the Arabian Peninsula*; Shield Area of Western Saudi Arabia (No. 560-A); US Geological Survey: Reston, VA, USA, 1989; 188p.

39. Al-Washmi, H.A. Sedimentological aspects and environmental conditions recognized from the bottom sediments of Al-Kharrar Lagoon, eastern Red Sea coastal plain, Saudi Arabia. *J. KAU Mar. Sci.* **1999**, *10*, 71–87. [CrossRef]

40. Al-Dubai, T.A.; Abu-Zied, R.H.; Basaham, A.S. Present environmental status of Al-Kharrar Lagoon, central of the eastern Red Sea coast, Saudi Arabia. *Arab. J. Geosci.* **2017**, *10*, 305. [CrossRef]

41. Aljahdali, M.O.; Alhassan, A.B. Ecological risk assessment of heavy metal contamination in mangrove habitats, using biochemical markers and pollution indices: A case study of Avicennia marina L. in the Rabigh lagoon, Red Sea. *Saudi J. Biol. Sci.* **2020**, *27*, 1174–1184. [CrossRef] [PubMed]

42. Bahrawi, J.A.; Elhag, M.; Aldhebiani, A.Y.; Galal, H.K.; Hegazy, A.K.; Alghailani, E. Soil erosion estimation using remote sensing techniques in Wadi Yalamlam Basin, Saudi Arabia. *Adv. Mater. Sci. Eng.* **2016**, *2016*, 1–8. [CrossRef]

43. Schmidt, G.L.; Jenkerson, C.B.; Masek, J.; Vermote, E.; Gao, F. Landsat Ecosystem Disturbance Adaptive Processing System (LEDAPS) Algorithm Description. Available online: https://pubs.er.usgs.gov/publication/ofr20131057 (accessed on 17 March 2015).

44. Vermote, E.; Justice, C.; Claverie, M.; Franch, B. Preliminary analysis of the performance of the Landsat 8/OLI land surface reflectance product. *Remote Sens. Environ.* **2016**, *185*, 46–56. [CrossRef]

45. Wilson, E.H.; Sader, S.A. Detection of forest harvest type using multiple dates of Landsat TM imagery. *Remote Sens. Environ.* **2002**, *80*, 385–396. [CrossRef]

46. Qi, J.; Chehbouni, A.; Huete, A.R.; Kerr, Y.H.; Sorooshian, S. A modified soil adjusted vegetation index. *Remote Sens. Environ.* **1994**, *48*, 119–126. [CrossRef]

47. Fensholt, R.; Horion, S.; Tagesson, T.; Ehammer, A.; Grogan, K.; Tian, F.; Huber, S.; Verbesselt, J.; Prince, S.D.; Tucker, C.J. Assessment of vegetation trends in drylands from time series of earth observation data. In *Remote Sensing Time Series*; Springer: Berlin/Heidelberg, Germany, 2015; pp. 159–182.

48. Ollinger, S.V. Sources of variability in canopy reflectance and the convergent properties of plants. *New Phytol.* **2011**, *189*, 375–394. [CrossRef]

49. Forkel, M.; Carvalhais, N.; Verbesselt, J.; Mahecha, M.D.; Neigh, C.S.R.; Reichstein, M. Trend change detection in NDVI time series: Effects of inter-annual variability and methodology. *Remote Sens.* **2013**, *5*, 2113–2144. [CrossRef]

50. Zeileis, A.; Shah, A.; Patnaik, I. Testing, monitoring, and dating structural changes in exchange rate regimes. *Comput. Stat. Data Anal.* **2010**, *54*, 1696–1706. [CrossRef]

51. Bai, J.; Perron, P. Computation and analysis of multiple structural change models. *J. Appl. Econom.* **2003**, *18*, 1–22. [CrossRef]

52. Jin, S.; Sader, S.A. Comparison of time series tasseled cap wetness and the normalized difference moisture index in detecting forest disturbances. *Remote Sens. Environ.* **2005**, *94*, 364–372. [CrossRef]

53. Lugo, A.E.; Snedaker, S.C. The ecology of mangroves. *Annu. Rev. Ecol. Syst.* **1974**, *5*, 39–64. [CrossRef]

54. Youssef, M.; El-Sorogy, A. Environmental assessment of heavy metal contamination in bottom sediments of Al-Kharrar lagoon, Rabigh, Red Sea, Saudi Arabia. *Arab. J. Geosci.* **2016**, *9*, 474. [CrossRef]

55. Yahiya, A.B. Environmental degradation and its impact on tourism in Jazan, KSA using remote sensing and GIS. *Int. J. Environ. Sci.* **2012**, *3*, 421–432.

56. Mayr, S.; Kuenzer, C.; Gessner, U.; Klein, I.; Rutzinger, M. Validation of Earth Observation Time-Series: A Review for Large-Area and Temporally Dense Land Surface Products. *Remote Sens.* **2019**, *11*, 2616. [CrossRef]

57. Krishnamurthy, P.; Mohanty, B.; Wijaya, E.; Lee, D.-Y.; Lim, T.-M.; Lin, Q.; Xu, J.; Loh, C.-S.; Kumar, P.P. Transcriptomics analysis of salt stress tolerance in the roots of the mangrove Avicennia officinalis. *Sci. Rep.* **2017**, *7*, 1–19. [CrossRef]

58. Vovides, A.C.; López-Portillo, J.; Bashan, Y. N 2-fixation along a gradient of long-term disturbance in tropical mangroves bordering the Gulf of Mexico. *Biol. Fertil. Soils* **2011**, *47*, 567–576. [CrossRef]

59. Twilley, R.R.; Rivera-Monroy, V.H.; Chen, R.; Botero, L. Adapting an ecological mangrove model to simulate trajectories in restoration ecology. *Mar. Pollut. Bull.* **1999**, *37*, 404–419. [CrossRef]

60. Reef, R.; Lovelock, C.E. Regulation of water balance in mangroves. *Ann. Bot.* **2015**, *115*, 385–395. [CrossRef]

forests

MDPI

Article

Monitoring Carbon Stock and Land-Use Change in 5000-Year-Old Juniper Forest Stand of Ziarat, Balochistan, through a Synergistic Approach

Hamayoon Jallat [1], Muhammad Fahim Khokhar [1,*], Kamziah Abdul Kudus [2], Mohd Nazre [2], Najam u Saqib [1], Usman Tahir [3] and Waseem Razzaq Khan [2,4,*]

[1] Institute of Environmental Science and Engineering, National University of Science and Technology, Islamabad 44000, Pakistan; hjallat.mses2017iese@student.nust.edu.pk (H.J.); nsaqib.mses2017iese@student.nust.edu.pk (N.u.S.)
[2] Department of Forest Science & Biodiversity, Faculty of Forestry and Environment, Universiti Putra Malaysia, UPM Serdang, Selangor 43400, Malaysia; kamziah@upm.edu.my (K.A.K.); nazre@upm.edu.my (M.N.)
[3] Department of Forest Sciences, Chair of Tropical and International Forestry, Faculty of Environmental Science, Technische Universität, 01069 Dresden, Germany; usman.tahir@mailbox.tu-dresden.de
[4] Institut Ecosains Borneo, Universiti Putra Malaysia, Bintulu Kampus, Sarawak 97008, Malaysia
* Correspondence: fahim.khokhar@iese.nust.edu.pk (M.F.K.); khanwaseem@upm.edu.my (W.R.K.); Tel.: +92-51-9085-4308 (M.F.K.); +60-19-3557-839 (W.R.K.)

Abstract: The Juniper forest reserve of Ziarat is one of the biggest *Juniperus* forests in the world. This study assessed the land-use changes and carbon stock of Ziarat. Different types of carbon pools were quantified in terms of storage in the study area in tons/ha i.e., above ground, soil, shrubs and litter. The Juniper species of this forest is putatively called *Juniperus excelsa* Beiberstein. To estimate above-ground biomass, different allometric equations were applied. Average above ground carbon stock of the forest was estimated as 8.34 ton/ha, 7.79 ton/ha and 8.4 ton/ha using each equation. Average carbon stock in soil, shrubs and litter was calculated as 24.35 ton/ha, 0.05 ton/ha and 1.52 ton/ha, respectively. Based on our results, soil carbon stock in the Juniper forest of Ziarat came out to be higher than the living biomass. Furthermore, the spatio-temporal classified maps for Ziarat showed that forest area has significantly decreased, while agricultural and barren lands increased from 1988 to 2018. This was supported by the fact that estimated carbon stock also showed a decreasing pattern between the evaluation periods of 1988 to 2018. Furthermore, the trend for land use and carbon stock was estimated post 2018 using a linear prediction model. The results corroborate the assumption that under a business as usual scenario, it is highly likely that the *Juniperus* forest will severely decline.

Keywords: Juniperus; above-ground biomass; land-use; allometric equation; satellite remote sensing; land cover classification

Citation: Jallat, H.; Khokhar, M.F.; Kudus, K.A.; Nazre, M.; Saqib, N.u; Tahir, U.; Khan, W.R. Monitoring Carbon Stock and Land-Use Change in 5000-Year-Old Juniper Forest Stand of Ziarat, Balochistan, through a Synergistic Approach. *Forests* **2021**, *12*, 51. https://doi.org/10.3390/f12010051

Received: 24 September 2020
Accepted: 17 November 2020
Published: 1 January 2021

Publisher's Note: MDPI stays neutral with regard to jurisdictional claims in published maps and institutional affiliations.

1. Introduction

The present forest area of the world is 4.06 billion ha and it has lost 178 million ha of forest since 1990, presenting a decrease of 4.2% [1]. It is estimated that the world's forest area has decreased from 31.6% to 31% of the total Earth's land area. With approximately 5% of the land area under forest, Pakistan finds its place among the low-forested countries in the world [2–4]. Major forest types in Pakistan include coastal mangroves, riverine forests, sub-tropical scrub forests, moist temperate conifer forests, dry temperate conifer forests, and irrigated plantations including linear plantations [5]. Based on the 'Forestry Sector Master Plan of 1992', 4,200,000 ha (4.8%) of Pakistan has natural forest [6]. Compared to that, in 2015, FAO (Food and Agriculture Organization) reported 1.9% or about 1,472,000 ha of forest area in Pakistan. This translates to a loss 2,728,000 ha of the forest cover, which has also been documented by the Pakistan Forestry Outlook study in 2009 and the Forest Resource Assessment in 2015. Evidently, there is a declining trend in the forest cover while agricultural and urban areas are expanding [7].

The genus *Juniperus* is one of the biggest conifer genera with 52 species worldwide and is extensively spread over the northern temperate region [8]. Juniper forest of Ziarat, also known as the second-largest forest of juniper in the world, that covers an area of about 110,000 hectares and was declared a Biosphere Reserve in 2013 [9]. The area of this Juniper forest as per the working plan 1960 is 247,166 acres (100,025 ha) [10]. However, Akram et al. [11] calculation based on object-based image analysis, found that the area of the juniper forest of Ziarat is 53,092 ha in 2010. This discrepancy clearly shows that the juniper forest of Ziarat faced the threat of both deforestation and forest degradation [12]. Not many studies on carbon stock assessment have been carried out in *Juniperus* forests across the globe and none have been conducted previously in the juniper forest of Ziarat. In the pinyon-juniper of western Colorado Plateau, mean above-ground woody carbon was estimated to be 5.2 ± 2.0 Mg C/ha [13]. In Gilgit Baltistan, Ismail et al. [14] studied the carbon stock of *Juniperus communis* using allometric equation of Jenkin et al. [15] and estimated the amount of carbon to be 1.96 ton/ha. Soils in juniper forests are also considered as a cost-effective carbon sink and conserving this type of forest is imperative for carbon sequestration [16].

The second-largest anthropogenic CO_2 emissions are from land-use changes such as rigorous cultivation, deforestation and logging [17]. In the past decades there has been an alarming rise in urbanization, illegal harvesting and agricultural activities [18]. Monitoring land-use and land-cover (LULC) change may help environmentalists, organizations and government offices to devise conservation strategies and management plans [19]. In addition, it could be a significant step to preserve and enhance carbon storage, if the land is properly managed, deforestation is controlled, and reforestation of the degraded land is carried out [20]. Above-ground biomass of forest plays a pivotal role as a terrestrial carbon sink in global carbon cycling [21]. As a carbon sink, forests are an economic treasure worth billions of dollars which currently absorb 30% of all the carbon dioxide emission globally every year [22].

Changes in land use such as converting land for agriculture purposes, results in lowering the soil organic carbon (SOC) or simply depleting carbon in the soil [23]. The conversion of rangeland into agricultural land reduces soil carbon [24]. The world's soil stores significantly much more carbon than the Earth's atmosphere [25]. The loss of soil carbon to the atmosphere may intensify the warming of the planet [26]. The carbon stock in soil has been greatly lost or widely degraded [27]. However, if good management practices are put in place, the SOC levels of the soil may be elevated along with enhancing soil quality [28]. Evidence suggests that SOC is affected by tree species and that some trees may be better at sequestering carbon in the soil [29].

Remote sensing (RS) and geographic information systems (GIS) serve as efficient tools for general and more detailed LULC change analysis [19]. LULC classification is an important research area in remote sensing. It is an established methodology in producing accurate, reliable and updated maps that are significant for ecological monitoring and management [30].

The two main objectives of this study are (1) to find the total loss in forest area of Ziarat over the past three decades using optical satellite imagery (2) to measure the biomass and estimate carbon stock in juniper forest of Ziarat and assess changes in carbon stock.

2. Materials and Methods

2.1. Site Description

Ziarat district lies in one of the six divisions of Balochistan province and happens to have one of the second-biggest reserves of juniper forest in the world. Juniper forest is unevenly spread between the two tehsils of Ziarat District, which are Sanjavi Tehsil and Ziarat Tehsil. Figure 1 shows the study area map of district Ziarat Balochistan.

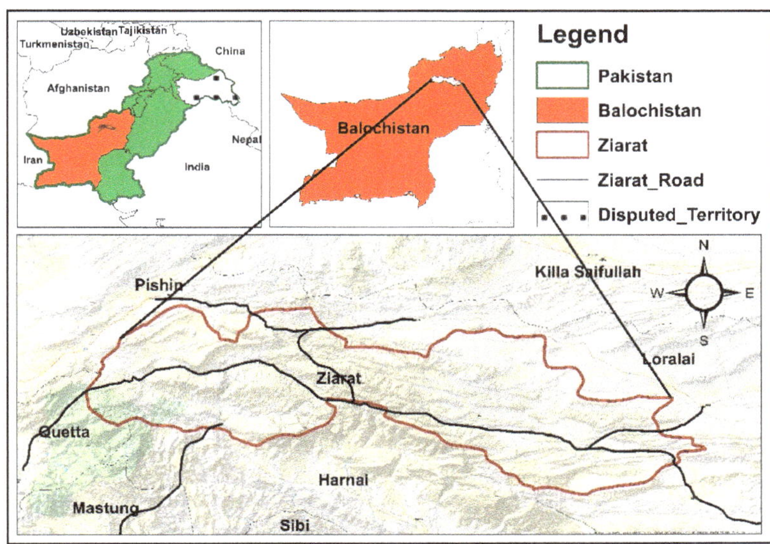

Figure 1. Study area showing district Ziarat Balochistan.

2.2. Monitoring Land-Use and Land-Cover (LULC) Changes

The archived Landsat imagery provide a unique opportunity to identify land use changes over the past 30 years [31]. Because of that, for the detection of the land cover for District Ziarat, images of Landsat 4-5 TM (Thematic Mapper) and Landsat 8 OLI (Operational Land Imager) with no or minimum cloud cover (0–10%) were selected and downloaded for the years of 1988, 1998, 2008, and 2018 from the USGS (United States Geological Survey) Earth Explorer website (https://earthexplorer.usgs.gov/). Two tiles (path row number 152/39 and 153/39) were downloaded for each period to cover the study area. The downloaded images belong to the month of September (Table 1) since it was found very feasible with almost no clouds present.

Table 1. Landsat images used for land cover classification.

Sensor	Year	Date	Bands (µm)
Landsat 8 OLI	2018	10-September 17-September	Band 2 (0.45–0.51) Band 3 (0.53–0.59) Band 4 (0.64–0.67) Band 5 (0.85–0.88) Band 6 (1.57–1.65) Band 7 (2.11–2.29)
Landsat 5 TM	2008	30-September 21-September	Band 1 (0.45–0.52) Band 2 (0.52–0.60)
	1998	03-September 10-September	Band 3 (0.63–0.69) Band 4 (0.77–0.90)
	1988	07-September 14-September	Band 5 (1.55–1.75) Band 7 (2.09–2.35)

The images were automatically atmospherically corrected using Semi-Automated Classification plugin in QGIS 3.6.2 (QGIS Development Team 2002). Images along with the metadata file were uploaded in the pre-processing Table DOS1 (Dark Object Subtraction) algorithm option was checked okay before running the processing chain. Atmospherically corrected images were opened in Arc-Map 10.3 (Esri 2014, Redlands, CA, USA), where the bands were stacked using the composite band tab in the Image Analysis window. The stacked images were mosaiced followed by extraction of study area using extract by mask algorithm. The study area was thoroughly examined on Google Earth maps. Both true and false color of Landsat images were observed to ensure that the visual interpretation was done correctly. Once the area was studied, pixels were assigned to the classes and classification was carried out via maximum likelihood classification. The area for each class was calculated in ArcMap and the data were further analyzed by using MS Excel. Line graphs and pie charts were formulated for easy analysis of the data.

2.3. Carbon Stock Assessment

This study estimated carbon from all the pools namely above ground tree biomass, shrubs, litter and soil C. Below ground biomass, however, was not considered in this study due to limited resources.

The field work was carried out in the month of February 2018. Six random clusters were selected throughout the study area, each with one primary and four secondary sampling units. The number of total plots were 30, however, due to absence of trees in some areas, the number of plots were reduced to 21 plots as depicted in Figure 2 below.

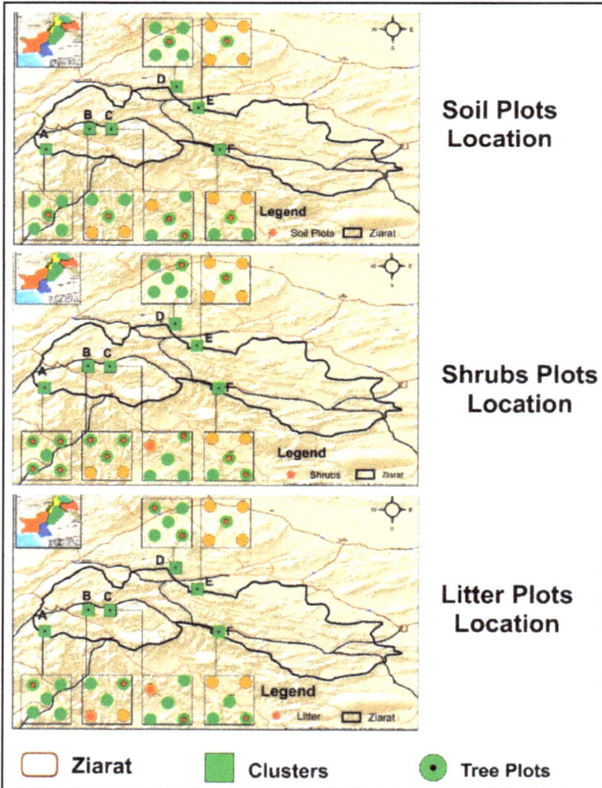

Figure 2. Location of tree plots, soil samples, shrubs samples and litter samples in the study area.

The above-ground biomass was calculated for all the plots, however, only one soil sample was collected from the primary plot of each cluster thus giving us six soil samples. Plots for above-ground biomass, shrubs, litter and soil had a radius of 17.8 m, 5.64 m, 0.56 m respectively, keeping in mind that soil and litter plots have the same radius. Shrubs and litter samples were collected as per their availability, and their location data points are displayed in Figure 2.

Tree height was measured using a clinometer and diameter was measured at breast height of 1.3 m using a caliper. All these data including elevation, site coordinates, time and date of sampling were also recorded. Living biomass data was entered into excel sheets to run the allometric equations. The allometric equations (see Table 2) used for biomass estimation were taken from Jenkin et al. [15], Ali [32] and Chave et al. [33]. The Jenkin et al. [15] equation has been used by the forest department of Gilgit Baltistan [14]. The equation used by Ali [32] and Chave et al. [33] has also been used by WWF (World Wide Fund) for the National Forest Inventory of Pakistan.

Table 2. Allometric equation used in the study.

Jenkin et al. (2003) Equation	Ali. (2015) Equation	Chave et al. (2005) Equation
$EXP(-0.7152 + 1.7029 \times LN(DBH))$	$0.1645 \times (p \times D\hat{}2 \times H)\hat{}0.8586$	$EXP(-2.187 + 0.916 \times LN(WD \times DBH\hat{}2 \times H))$

EXP is the exponential, *LN* is the natural logarithm, *DBH* is the Diameter at Breast Height, *p* is the wood gravity, *D* is the Diameter, *WD* is the Wood Density, and *H* is the Height.

Fraction of carbon in the above ground living biomass can be assessed using the following equation adopted from FAO [34].

$$Total\ Carbon\ in\ above\ ground\ biomass\ (kg) = 0.475 \times Above\ ground\ Biomass\ (kg) \tag{1}$$

where the above ground biomass is assessed using the allometric equations listed in Table 2 The above ground living biomass volume formula was derived from the stem biomass formula of FAO [35] as mentioned below:

$$Biomass\ (kg) = V_s \times WD \times 1000 \tag{2}$$

where V_s is stem volume and *WD* is wood density. Similarly, tree volume can be calculated by using following equation:

$$Biomass\ (kg) = V_t \times WD \times 1000 \tag{3}$$

$$V_t = Biomass/WD \times 1000 \tag{4}$$

where V_t is tree volume and *WD* is wood density.

The carbon stock of each carbon pool in tons was converted into tons per hectare using the following formula:

$$C\ in\ ton/ha_{pool} = C\ in\ tons/Area\ of\ the\ plot_{pool} \tag{5}$$

Total above ground carbon (million tons) for the total forest area, in each respective year of 1988, 1998, 2008 and 2018 was estimated using the following equation:

$$Total\ forest\ Above\ Ground\ Carbon\ (million\ tons) = C\ in\ ton/ha \times Area\ of\ forest\ in\ ha\ for\ each\ year \tag{6}$$

Total forest carbon for the years before 2018 were estimated using average carbon stock of the three allometric equations used. Additionally, total above carbon for the year 2028, 2038 and 2048 was also estimated using the above formula. Area in hectares of these future years were inferred using the linear forecast model in Microsoft excel worksheet 2013 (Supplementary Material). Total carbon of the forest (million tons) for each year of past and future is presented in results.

The method for estimating total carbon stock of the forest by taking the product of average carbon ha^{-1} and the total forest cover (ha) at a specific temporal period was also used by Mannan et al. [18].

2.4. Linear Forecast Model

Linear forecast function uses a linear regression method to predict future values based on historical figures. It is a method of defining the relationship between two or more variables in a way that changes in dependent variable can be accounted by changes in the independent variable. In our study carbon stock and forest area are the dependent variables and time (against year) is the independent variable.

2.5. Soil Carbon Stocks

Soil samples were collected at three depths; 0 to10 cm, 10 to 20 cm and 20 to 30 cm. An auger was used for the collection of soil samples. A total of Six soil samples were collected, one from each cluster. Bulk density of each soil sample was calculated in g/cm^3. For the determination of carbon in soil, the Walkley–Black titration procedure was applied. To find soil C in grams, the following equation was utilized.

$$SC \ (g) = BD \ g/cm^3 \times SOC \ (\%) \times HT \ cm \times 100 \tag{7}$$

where *SC* Soil Carbon, *BD* is the Bulk Density, *SOC* is the Soil Organic Carbon, and *HT* is the Horizon Thickness.

Soil carbon in grams was converted into tons and further into ton/hectare by dividing it by the plot area.

3. Results

3.1. LULC Changes 1988–2018

Land cover maps (Figure 3) showed drastic changes in the land use pattern of Ziarat District. Forest area, which is the major focus of the study decreased from 21.5% in 1988 to 15.5% in 2018 showing a decrease of 6% in the total area. This represents significant deforestation over the past decades. Similarly, agriculture area increased from 1.5% in 1988 to 3.5% in 2018. This might be due to an increasing population and demand for agricultural production (Figure 3). As per the housing and population census, the population of Ziarat increased from 32,196 people in 1981 to 160,422 people in 2017. Barren land also significantly increased from 76.5% in 1988 to 81% in 2018 due to deforestation.

As shown in Figure 4 (land use change trend and forecast), the forest area decreased from 71,005 hectares in 1988 to 50,311 hectares in 2018, depicting a loss through deforestation of 20,694 hectares in the past 30 years. On the other hand, the agriculture land increased from 4848 hectares in 1988 to 11,625 hectares in 2018 representing an increase of 6777 hectares.

Figure 4 also shows the linear forecast values inferred for next 30 years. The forecast analysis shows that the forest area will decrease to 29,976.5 hectares in 2048, while the agricultural and the barren land area will increase to 17,826.7 hectares and 279,259 hectares, respectively, in the coming three decades.

Table 3 below shows the accuracy assessment of the land-use maps of Ziarat district. Accuracy assessment was performed in Arc Map using the confusion matrix, where 140 training points were selected for each temporal map. The confusion matrix shows producer's and user's accuracy for each land-use category of the respective yearly maps. Accuracy from the perspective of the map designer is termed as the producer's accuracy and it shows the probability of whether land cover has been classified correctly. User's accuracy tells us the correctness of the map from a user's viewpoint. It tells us the probability of how often the classified class on the map will be present on the ground [36,37]. It is quite high and above 90%. The overall accuracy of the classification for 1988, 1998, 2008 and 2018 was 0.97, 0.98, 0.98, and 0.97. Kappa analysis was undertaken if the performance of the classification did well in comparison to randomly assigning values. This ranged from -1 to 1 and a value of zero represents that the classification was no better random value assignment [36,37]. The Kappa coefficient was 0.96, 0.98, 0.97 and 0.96, respectively, for each consecutive year.

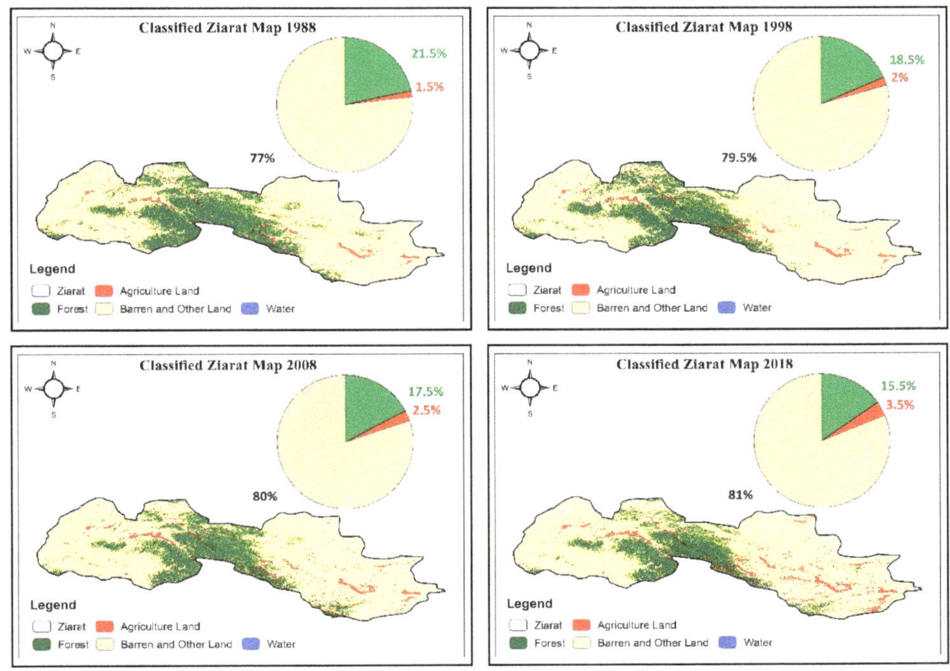

Figure 3. Land-use and land-cover (LULC) maps of the study area for the year 1988, 1998, 2008 and 2018.

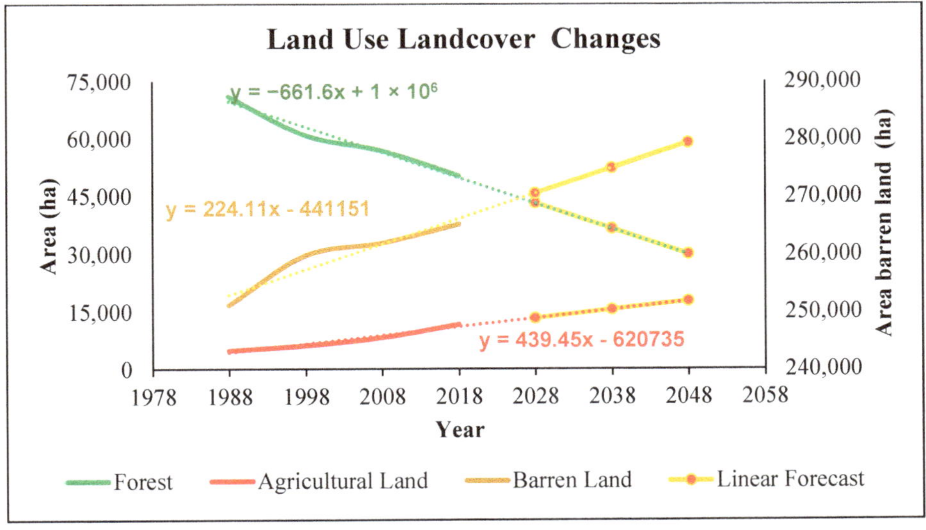

Figure 4. Land-use change trend (1988–2018) with inferred values for future 30 years using linear forecast.

Table 3. Accuracy assessment of land use maps.

	1988		1998		2008		2018	
	Producer's Accuracy %	User's Accuracy %	Producer's Accuracy%	User's Accuracy%	Producer's Accuracy %	User's Accuracy %	Producer's Accuracy %	User's Accuracy %
Forest	92	100	100	100	94	100	92	100
Agriculture	100	100	100	100	100	100	100	100
Barren and Other land	100	92.59	100	100	100	94.33	100	92.59
Water	100	100	80	100	100	100	100	100
Kappa Coefficient	0.96		0.98		0.97		0.96	
Overall Accuracy	0.97		0.98		0.98		0.97	

3.2. Carbon Stock Assessment

Table 4 shows plot-wise estimates of different parameters of the juniper tree namely the number of trees, average height, average diameter and basal area. The maximum number of trees was recorded in plot no 14 which is 58 trees, and the minimum number of trees was in plot 21 with 10 individuals. The maximum average height and diameter were in plot 18 and the minimum average height and diameter were in plot 15. The maximum and minimum average height is 5.81 and 4.38 m, respectively. The highest and lowest diameter came out to be 23.94 and 10.01 cm. The maximum basal area was in plot 17 of 2.25 m^2. The minimum basal area was 0.19 m^2 in plot 21.

Table 4. Parameter estimation of sampled plots.

Plot ID	Number of Trees	Average Height (m)	Average Diameter (cm)	Basal Area (m^2)
Plot 1	14	4.96	18.27	0.72
Plot 2	20	4.65	13.71	0.48
Plot 3	30	5.01	17.33	1.36
Plot 4	22	4.60	11.22	0.25
Plot 5	26	4.84	12.80	0.40
Plot 6	24	5.26	17.91	0.85
Plot 7	20	4.62	12.43	0.36
Plot 8	24	5.16	16.55	0.67
Plot 9	46	4.64	12.50	0.86
Plot 10	23	5.13	18.43	0.97
Plot 11	16	4.75	12.29	0.22
Plot 12	21	4.79	12.84	0.33
Plot 13	38	4.89	14.36	0.87
Plot 14	**58**	4.72	13.27	1.29
Plot 15	55	**4.38**	**10.01**	0.55
Plot 16	21	4.44	11.83	0.36
Plot 17	50	5.17	18.35	**2.25**
Plot 18	16	**5.81**	**23.94**	0.98
Plot 19	21	5.61	20.66	0.86
Plot 20	35	4.98	16.73	1.23
Plot 21	**10**	4.95	14.2	**0.19**
		4.92	**15.22**	**0.76**

The bold in this graph depicts the maximum and minimum values in all the plots.

The overall average height of all the plots was 4.92 m while the overall average diameter was 15.22 cm. The overall average basal area was 0.76 m^2.

Table 5 below shows the total biomass and the total volume of juniper forest using three different above-ground allometric equations. Total biomass calculated from all three equations were 37,281.51 kg, 24,812.12 kg and 37,518.67 kg respectively. The volume estimated using these equations were 74.56 m^3, 69.62 m^3 and 75.04 m^3. It can be noted that there was a small difference between the values of the equations of Jenkin et al. [15] and Chave et al. [33]. But there was a visible difference in the biomass and volume calculated using the equation of Ali [32].

Table 5. Comparison of total biomass and total volume of the three respective allometric equations.

	Jenkin et al. (2003)		Ali (2015)		Chave et al. (2005)	
Serial no	Total Biomass kg	Total Volume m^3	Total Biomass kg	Total Volume m^3	Total Biomass kg	Total Volume m^3
Plot 1	1469.48	2.94	1525.63	3.05	1733.31	3.47
Plot 2	1136.97	2.27	1042.32	2.08	1108.32	2.22
Plot 3	2808.46	5.62	2882.99	5.77	3273.74	6.55
Plot 4	726.67	1.45	564.27	1.13	551.60	1.10
Plot 5	1092.21	2.18	894.63	1.79	898.08	1.80
Plot 6	1969.00	3.94	1857.40	3.71	1998.63	4.00
Plot 7	900.20	1.80	779.92	1.56	807.14	1.61
Plot 8	1637.96	3.28	1471.67	2.94	1539.76	3.08
Plot 9	2121.05	4.24	1868.06	3.74	1959.28	3.92
Plot 10	2148.44	4.30	2109.62	4.22	2317.58	4.64
Plot 11	589.45	1.18	476.99	0.95	475.87	0.95
Plot 12	889.08	1.78	725.82	1.45	725.00	1.45
Plot 13	2139.73	4.28	1907.67	3.82	1999.62	4.00
Plot 14	3057.40	6.11	2791.66	5.58	2988.37	5.98
Plot 15	1567.66	3.14	1217.62	2.44	1200.58	2.40
Plot 16	826.38	1.65	778.57	1.56	851.04	1.70
Plot 17	4782.44	9.56	4805.25	9.61	5381.93	10.76
Plot 18	2107.65	4.22	2115.15	4.23	2346.73	4.69
Plot 19	2021.00	4.04	1893.39	3.79	2019.08	4.04
Plot 20	2786.63	5.57	2679.93	5.36	2914.52	5.83
Plot 21	503.65	1.01	423.55	0.85	428.50	0.86
Total	**37,281.51**	**74.56**	**34,812.12**	**69.62**	**37,518.67**	**75.04**

Figure 5 represents the correlation between the parameters of tree as specified in Tables 3 and 4. There is a good correlation between the height and diameter I-e 0.89, and this does not vary among the allometric equation. This shows that the height of the tree increases with the increase in diameter of the juniper tree.

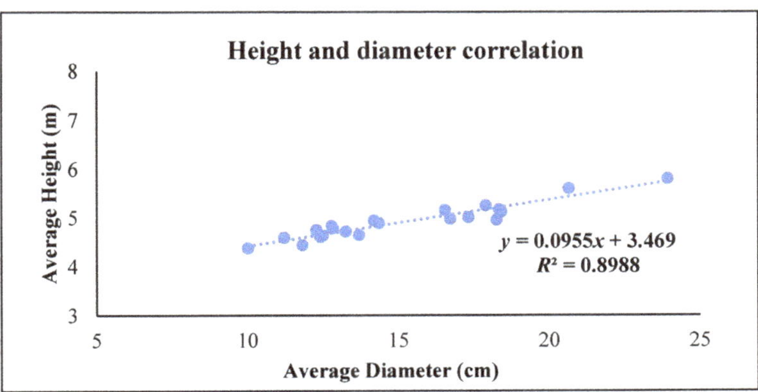

Figure 5. Correlation of average height and average diameter from the sampled plots.

Table 6 shows an overall plot summary of carbon stock in above ground pool using the three equations as mentioned earlier. Plot 17 contains the highest amount of above ground biomass i.e., 22.48, 22.58 and 25.30 ton/ha, and the lowest above ground biomass is found in plot 21 i.e., 2.37, 1.99, and 2.01 ton/ha. The overall average living biomass of the forest of all plots using the three equations is 8.34, 7.79 and 8.4 ton/ha, respectively.

Table 6. Comparison of the above ground carbon using the allometric equations from Jenkin et al. [15], Ali [32], and Chave et al. [33].

Serial. No	ABG (t/ha) Using Jenkin et al. (2003) Allometric Equation	ABG (t/ha) Using Ali (2015) Allometric Equation	ABG (t/ha) Using Chave et al. (2005) Allometric Equation
Plot 1	6.91	7.17	8.15
Plot 2	5.34	4.90	5.21
Plot 3	13.20	13.55	15.39
Plot 4	3.42	2.65	2.59
Plot 5	5.13	4.20	4.22
Plot 6	9.25	8.73	9.39
Plot 7	4.23	3.67	3.79
Plot 8	7.70	6.92	7.24
Plot 9	9.97	8.78	9.21
Plot 10	10.10	9.92	10.89
Plot 11	2.77	2.24	2.24
Plot 12	4.18	3.41	3.41
Plot 13	10.06	8.97	9.40
Plot 14	14.37	13.12	14.05
Plot 15	7.37	5.72	5.64
Plot 16	3.88	3.66	4.00
Plot 17	22.48	22.58	25.30
Plot 18	9.91	9.94	11.03
Plot 19	9.50	8.90	9.49
Plot 20	13.10	12.60	13.70
Plot 21	2.37	1.99	2.01
Average	**8.34**	**7.79**	**8.40**

Table 7 below shows the carbon estimation of the pools: soil, shrubs and litter. The respective plots where there is highest and lowest carbon stock have been highlighted. The overall average soil carbon of the juniper forest is 24.35 ton/ha. The overall carbon stored in shrubs and litter is 4.66 and 1.52 ton/ha, respectively.

Figure 6 below shows the total biomass of the juniper forest of Ziarat since 1988 using three different equations. The initial bars on the left of the graph show the total estimated carbon stock of the forest for the years 1988, 1998, 2008 and 2018. The bars on the right represent the inferred values of the total forest carbon stock for the year 2028, 2038 and 2048.

The Jenkin et al. [15] equation in blue bars estimates total above ground carbon stock of the forest in 1988 as 0.59 million tons. It reduced to 0.51 million tons, 0.47 million tons and 0.42 million tons in 1998, 2008, and 2018, respectively. The forecast values using the same equation shows that the total carbon of the forest may reduce to 0.36 million tons in 2028, 0.31 million tons in 2038 and 0.25 million tons in 2048.

Similarly, according to the Ali [32] allometric equation estimates, presented in the orange bar (Figure 6), the total above ground carbon stock of the forest in 1988 was 0.55 million tons. It reduced to 0.47 million tons, 0.44 million tons, and 0.39 million tons in 1998, 2008 and 2018, respectively. The forecast values show that the total carbon of the juniper forest may reduce to 0.34 million tons in 2028, 0.29 million tons in 2038 and 0.23 million tons in 2048.

Moreover, the estimates of the Chave et al. [33] allometric equation, displayed as green bars, are also interesting and almost like the Jenkin et al. 2003 allometric equation. (Figure 6). The total above ground carbon stock of the forest in 1988 was 0.60 million tons. It reduced to 0.51 million tons, 0.48 million tons and 0.42 million tons in 1998, 2008 and 2018, respectively. The forecast values show that the total carbon of the juniper forest may reduce to 0.36 million tons in 2028, 0.31 million tons in 2038 and 0.25 million tons in 2048.

Table 7. Estimated carbon in soil, shrubs and litter in sampled plots.

Serial. No	Soil Carbon ton/ha	Shrubs ton/ha	Litter ton/ha
Plot 1		0.057	
Plot 2		0.045	2.22
Plot 3	**28.06**	0.033	
Plot 4		0.042	
Plot 5			
Plot 6		0.033	**3.83**
Plot 7	27.22	0.075	
Plot 8		0.016	1.87
Plot 9			
Plot 10		0.085	**1.37**
Plot 11	27.54	**0.102**	
Plot 12			1.23
Plot 13			
Plot 14			0.55
Plot 15	**17.76**	0.066	
Plot 16			0.45
Plot 17		0.047	
Plot 18	22.53	**0.014**	0.55
Plot 19			
Plot 20	22.99	0.019	1.62
Plot 21		0.018	
Average	**24.35**	**0.05**	**1.52**

The bold values are the high and the low values in all the data set.

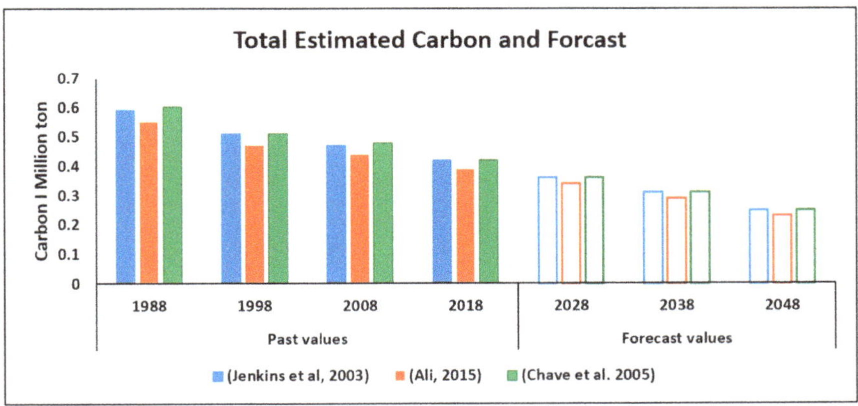

Figure 6. Total forest carbon in million tons using the equations of Jenkin et al. 2003, Ali (2015) and Chave et al., 2005 from 1988 to 2018 and for the future 30 years using linear forecast analysis.

4. Discussion

Ziarat juniper forest was declared a biosphere reserve in 2013. Such forests will play a pivotal role in future in furthering the cause of carbon sequestration [10]. However, they face a visible threat of deforestation and forest degradation [38]. Major drivers of deforestation and forest degradation are population expansion, agriculture intensification, fuel wood consumption, poor regeneration, illegal cutting, overgrazing, canopy dieback, mistletoe attack, periodic drought and medicinal use of the juniper tree [38,39]. For identifying deforestation, a quantitative assessment of the land-use change was performed for four land use classes namely, forest, agriculture, water and barren land. Focus of the study was reduction in forest cover, which according to our results was accounted for by the increase in agricultural area.

There is an inverse relationship between forest resources and both population as well as the amount of land used for agriculture. With population increase, more and more forests are encroached upon for harvesting firewood and timber, thereby increasing the non-forest area in Ziarat district [40]. The population increase has led to the increase in dependency on agriculture [12], thus having a synergetic effect on the forest. Besides other drivers, *Archeothobium oxycedri*, known as dwarf mistletoe, also damages the *Juniperus* species [41]. Sarangzai et al. [42] also reported the spread of the same parasitic and infectious plant in the juniper forest of Ziarat. Harvesting fuelwood is also one of the major drivers of deforestation since there is no other source of fuel to keep the local people warm in winter [43]. Even though natural gas has been provided to Ziarat, the pressure in gas pipelines is extremely low thus compelling residents to cut trees in the winter season [44]. Most of the juniper forest have low seedlings and show no adequate regeneration thus slowing down recovery time [45].

Urban areas were not classified, because most of the construction sites are either muddy houses or wooden-built sites, thus making it extremely difficult to separate its spectral information in Landsat imagery. The area also has myriad soil forms and colors resulting in the overlapping of the pixels during classification. Even the forest area was difficult to classify since the juniper forest is a sparse forest with less tree density. However, it was classified very well when observed simultaneously with Google Earth imagery. The results of this study are also comparable to the results of a study conducted by WWF and SUPPARCO (Pakistan Space & Upper Atmosphere Research Commission) in 2012 on the Juniper Forest of Ziarat using SPOT data [11]. This suggests that the Landsat data are good for forest classification and land-use change.

Ismail et al. [14] estimated the above ground carbon per hectare of *Juniperus communis*. The total number of juniper trees counted were 278 and the estimated average carbon was 1.96 ton/ha with a total basal area of 12.28 m^2. This value is substantially less compared to our estimated carbon of 8.34 ton/ha and a total basal area of 16.06 m^2 using the same equation for *Juniperus excelsa*.

Since no specific equations exist for the *Juniperus excelsa* species, it is recommended that the forest department in collaboration with academia develop an allometric equation for it. If developing an equation is not possible or not within the capacity, it is recommended that the three equations used in the study are used for any future carbon stock study. The three equations used in this study gave very similar results.

The data were collected from 21 plots having 585 trees. The results show that the soil contains more carbon than trees, which may be attributed to the compact soil of the area, low temperature conditions and the less dense/sparse nature of the forest. It may also be due to the age of the forest, which is very old thus, accumulating humus for thousands of years. One of the major reasons for less biomass in the tree is small height and the very sparse nature of the forest reserve.

The juniper forest of Ziarat is an extremely rare forest and needs to be protected and sustained. The forest area has decreased from 10,025 hectares in 1960 to 53,092 hectares in 2010 and is a clear manifestation of threats faced by the juniper forest of Ziarat, Balochistan. [5,11]. This comprises multiple factors. The first one is the population of the area that has dramatically increased (about 400%) from 1981 to 2017 as per the population and housing census. Besides marble mining at few places, there is no such industrial activity in the Ziarat district. Therefore, the population must depend on natural resources for their daily subsistence. The agriculture area has also increased (2% from 1988 to 2018) as indicated in the previous section of this study. Other factors include medicinal use of berries, climate change, timber extraction, tourism and poor forest management by the forest department. All these factors are clearly posing an enhanced threat to preservation of these ancient forests of Ziarat. Keeping the current scenario and past practices in view, it is most probable that the forest area will keep on shrinking as mentioned by various studies presented in Figure 6.

Moreover, the protection regime should not be only limited to Ziarat but also to the juniper forest resources in Loralai, Harnai, Quetta and Pishin. Despite all the services the juniper forest provides, it has not been taken care of in a viable way resulting in its deforestation and degradation. If a similar trend of disregard for this precious and ancient forest continues, we might only study about this archeological heritage in archives. For this purpose, the forest department must take practical steps for the conservation of the juniper forest of Balochistan.

5. Conclusions

The study concluded that the ancient juniper forest of Ziarat is an important carbon sink, storing a significant amount of carbon in all its pools i.e., above-ground live tree biomass, soil, shrubs and litter. The soil of the juniper forest stores more carbon than the living biomass due to low canopy cover and scattered growth of the trees. Similarly, the litter contains more carbon stock than the shrubs since the quantity of litter found in the plots was higher than the shrubs. Land-use maps showed a tremendous change in the forest cover of Ziarat Balochistan from 1988 to 2018 with a decreasing forest trend and increasing agricultural trend. Furthermore, carbon stock of juniper has also decreased over the past three decades due to excessive deforestation and forest degradation. Clearly, the juniper forest is threatened and if the same pace of deforestation continues, it is very likely that this forest will soon be wiped out. Additionally, the carbon stock of the juniper forest was assessed using three different equations which gave similar results, so it is suggested that these equations may be used for future carbon stock studies on the juniper forest. It is also concluded that, if no appropriate steps are taken towards the conservation of this forest, we may lose the ancient world biosphere reserve in a very short time.

Supplementary Materials: The following are available online at https://www.mdpi.com/1999-490 7/12/1/51/s1, Table S1: Linear forecast model.

Author Contributions: Conceptualization, H.J., M.F.K. and W.R.K.; Methodology, H.J. and N.u.S.; Formal Analysis, K.A.K. and H.J; Data, H.J.; Writing Original Draft Preparation, H.J., W.R.K. and M.F.K.; Writing Review & Editing, M.N.; Visualization, U.T.; Project Administration, M.F.K.; Funding Acquisition, M.N. All authors have read and agreed to the published version of the manuscript.

Funding: The authors are grateful to Universiti Putra Malaysia (UPM) and NUST for providing financial support from the postgraduate RnD fund from S and T Mega Project to conduct this study.

Acknowledgments: The authors acknowledge all the members of research group (C-CARGO) and Universiti Putra Malaysia researchers for being supportive.

Conflicts of Interest: The authors declare no conflict of interest.

References

1. FAO. Food and Agriculture Organization, Global Forest Resources Assessment 2020–Key findings. Rome. Available online: https://doi.org/10.4060/ca8753en (accessed on 13 November 2020).
2. Khan, W.R.; Khokhar, M.F.; Sana, M.; Naila, Y.; Qurban, A.P.; Muhammad, N.R. Assessing the context of REDD+ in Murree hills forest of Pakistan. *Adv. Environ. Biol.* **2015**, *9*, 15–20.
3. Munawar, S.; Khokhar, F.; Atif, S. Reducing Emissions from Deforestation and Forest Degradation implementation in northern Pakistan. *Int. Biodeterior. Biodegrad.* **2015**, *102*, 316–323. [CrossRef]
4. Yasmin, N.; Khokhar, M.F.; Tanveer, S.; Saqib, Z.; Khan, W.R. Dynamical assessment of vegetation trends over Margalla Hills National park by using MODIS vegetation indices. *Pak. J. Agri. Sci.* **2016**, *53*, 4.
5. Forest Department. *Ziarat Forest Management Plan*; Government of Balochistan: Quetta, Pakistan, 1960.
6. MOCC. *National Forest Policy*; Government of Pakistan: Islamabad, Pakistan, 2015.
7. Forests & Ministry of Environment. *Pakistan Forestry Outlook Study*; Food and Agriculture Organization: Bangkok, Thailand, 2009.
8. Ali, S.; Ali, W.; Khan, S.; Khan, A.; Rahman, Z.U.; Iqbal, A. Forest cover change and carbon stock assessment in Swat valley using remote sensing and geographical information systems. *Pure Appl. Biol.* **2017**, *6*, 850–856. [CrossRef]
9. Farjon, A. The taxonomy of multiseed junipers (Juniperus Sect. Sabina) in southwest Asia and east Africa (Taxonomic notes on Cupressaceae I). *Edinb. J. Bot.* **1992**, *49*, 251–283. [CrossRef]
10. UNESCO. United Nations Educational, Scientific and Cultural Organization, Ziarat Juniper Forest. Available online: http://whc.unesco.org/en/tentativelists/6116/ (accessed on 21 March 2018).

11. Akram, U.; Shahzad, N.; Saeed, U.; Naeem, S.; Hashmi, S.G.M.; Iqbal, I.A. Juniper Forest Belt Assessment Using Object Based Image Analysis, Sulaiman Range, Pakistan. In Proceedings of the 7th International Conference on Cooperation and Promotion of Information Resources in Science and Technology, Nanjing, China, 22–25 November 2012.

12. Hosonuma, N.; Herold, M.; De Sy, V.; De Fries, R.S.; Brockhaus, M.; Verchot, L.; Angelsen, A.; Romijn, E. An assessment of deforestation and forest degradation drivers in developing countries. *Environ. Res. Lett.* **2012**, *7*, 044009. [CrossRef]

13. Huang, C.; Asner, G.P.; Martin, R.E.; Barger, N.N.; Neff, J.C. Multiscale analysis of tree cover and aboveground carbon stocks in pinyon–juniper woodlands. *Ecol. Appl.* **2009**, *19*, 668–681. [CrossRef] [PubMed]

14. Ismail, I.; Sohail, M.; Gilani, H.; Ali, A.; Hussain, K.; Hussain, K.; Karky, B.S.; Qamer, F.M.; Qazi, W.; Ning, W.; et al. Forest inventory and analysis in Gilgit-Baltistan: A contribution towards developing a forest inventory for all Pakistan. *Int. J. Clim. Chang. Strat. Manag.* **2018**, *10*, 616–631. [CrossRef]

15. Jenkins, J.C.; Chojnacky, D.C.; Heath, L.S.; Birdsey, R.A. National-scale biomass estimators for United States tree species. *For. Sci.* **2003**, *49*, 12–35.

16. Charro, E.; Moyano, A.; Cabezón, R. The potential of *Juniperus thurifera* to sequester carbon in semi-arid forest soil in Spain. *Forests* **2017**, *8*, 330. [CrossRef]

17. Le Quéré, C.; Raupach, M.R.; Canadell, J.G.; Marland, G.; Bopp, L.; Ciais, P.; Conway, T.J.; Doney, S.C.; Feely, R.A.; Foster, P.; et al. Trends in the sources and sinks of carbon dioxide. *Nat. Geosci.* **2009**, *2*, 831. [CrossRef]

18. Mannan, A.; Liu, J.; Zhongke, F.; Khan, T.U.; Saeed, S.; Mukete, B.; Chaoyong, S.; Yongxiang, F.; Ahmad, A.; Amir, M.; et al. Application of land-use/land cover changes in monitoring and projecting forest biomass carbon loss in Pakistan. *Glob. Ecol. Conserv.* **2019**, *17*, e00535. [CrossRef]

19. Gidado, K.; Khairul, M.; Kamarudin, M.K.A.; Firdaus, A.; Nalado, A.M.; Saudi, A.S.M.; Saad, M.H.M.; Ibrahim, S. Analysis of Spatiotemporal Land Use and Land Cover Changes using Remote Sensing and GIS: A Review. *Int. J. Eng. Technol.* **2018**, *7*, 159–162. [CrossRef]

20. Ahmad, A.; Nizami, S.M. Carbon stocks of different land uses in the Kumrat valley, Hindu Kush Region of Pakistan. *J. For. Res.* **2015**, *26*, 57–64. [CrossRef]

21. Vashum, K.T.; Jayakumar, S. Methods to estimate above-ground biomass and carbon stock in natural forests-A review. *J. Ecosyst. Echogr.* **2012**, *2*, 1–7. [CrossRef]

22. Canadell, J.G.; Raupach, M.R. Managing forests for climate change mitigation. *Science* **2008**, *320*, 1456–1457. [CrossRef]

23. Lal, R. Forest soils and carbon sequestration. *For. Ecol. Manag.* **2005**, *220*, 242–258. [CrossRef]

24. Shah, S.; Ahmad, A.; Khan, A. Soil organic carbon stock estimation in range lands of Kumrat Dir Kohistan KPK Pakistan. *J. Ecol. Nat. Environ.* **2015**, *7*, 277–288.

25. Davidson, E.A.; Janssens, I.A. Temperature sensitivity of soil carbon decomposition and feedbacks to climate change. *Nature* **2006**, *440*, 165. [CrossRef]

26. Bradford, M.A.; Wieder, W.R.; Bonan, G.B.; Fierer, N.; Raymond, P.A.; Crowther, T.W. Managing uncertainty in soil carbon feedbacks to climate change. *Nat. Clim. Chang.* **2016**, *6*, 751. [CrossRef]

27. Jackson, R.B.; Lajtha, K.; Crow, S.E.; Hugelius, G.; Kramer, M.G.; Piñeiro, G. The ecology of soil carbon: Pools, vulnerabilities, and biotic and abiotic controls. *Annu. Rev. Ecol. Evol. Syst.* **2017**, *48*, 419–445. [CrossRef]

28. Lal, R.; Negassa, W.; Lorenz, K. Carbon sequestration in soil. *Curr. Opin. Environ. Sustain.* **2015**, *15*, 79–86. [CrossRef]

29. Vesterdal, L.; Clarke, N.; Sigurdsson, B.D.; Gundersen, P. Do tree species influence soil carbon stocks in temperate and boreal forests? *For. Ecol. Manag.* **2013**, *309*, 4–18. [CrossRef]

30. Balcik, F.B.; Kuzucu, A.K. Determination of Land Cover/Land Use Using SPOT 7 Data With Supervised Classification Methods. In Proceedings of the 3rd International GeoAdvances Workshop, Istanbul, Turkey, 16–17 October 2016.

31. Huang, H.; Chen, Y.; Clinton, N.; Wang, J.; Wang, X.; Liu, C.; Zheng, Y. Mapping major land cover dynamics in Beijing using all Landsat images in Google Earth Engine. *Remote Sens. Environ.* **2017**, *202*, 166–176. [CrossRef]

32. Ali, A. *Biomass and Carbon Tables for Major Tree Species of Gilgit Baltistan, Pakistan*; Gilgit-Baltistan Forests, Wildlife and Environment Department: Gilgit, Pakistan, 2015.

33. Chave, J.; Andalo, C.; Brown, S.; Cairns, M.A.; Chambers, J.Q.; Eamus, D.; Fölster, H.; Fromard, F.; Higuchi, N.; Kira, T.; et al. Tree allometry and improved estimation of carbon stocks and balance in tropical forests. *Oecologia* **2005**, *145*, 87–99. [CrossRef] [PubMed]

34. FAO. Food and Agriculture Organization, Knowledge Reference for National Forest assessments—Modeling for Estimation and Monitoring. Available online: http://www.fao.org/forestry/17111/en/ (accessed on 13 November 2020).

35. FAO. Food and Agriculture Organization, Chapter 3, Assessment of Biomass and Carbon Stock in Present Land Use. Available online: http://www.fao.org/3/y5490e/y5490e07.htm#TopOfPage (accessed on 13 November 2020).

36. Story, M.; Congalton, R. Accuracy assessment: A user's perspective. *Photogramm. Eng. Remote Sens.* **1986**, *52*, 397–399.

37. Congalton, R. A review of assessing the accuracy of classifications of remotely sensed data. *Remote Sens. Environ.* **1991**, *37*, 35–46. [CrossRef]

38. Achakzai, A.K.K.; Batool, H.; Aqeel, T.; Bazai, Z.A. A comparative study of the deforestation and regeneration status of Ziarat Juniper forest. *Pak. J. Bot.* **2013**, *45*, 1169–1172.

39. Sarangzai, A.M.; Ahmed, M.; Ahmed, A.; Tareen, L.; Jan, S.U. The ecology and dynamics of Juniperus excelsa forest in Balochistan-Pakistan. *Pak. J. Bot.* **2012**, *44*, 1617–1625.

40. Gul, S.; Khan, M.A.; Khair, S.M. Population Increase: A Major Cause of Deforestation in District Ziarat. *J. Appl. Emerg. Sci.* **2016**, *5*, 124–132.
41. Ciesla, W.M.; Mohammed, G.; Buzdar, A.H. Juniper dwarf mistletoe, Arceuthobium oxycedri (DC.) M. Bieb, in Balochistan Province, Pakistan. *For. Chron.* **1998**, *74*, 549–553. [CrossRef]
42. Sarangzai, A.M.; Khan, N.A.; Wahab, M.; Kakar, A. New spread of dwarf mistletoe (*Arceuthobium oxycedri*) in juniper forests, Ziarat, Balochistan, Pakistan. *Pak. J. Bot.* **2010**, *42*, 3709–3714.
43. BBC—British Broadcasting Corporation. Pakistan's Ziarat: An Ancient Juniper Forest and Its Living Fossils. Available online: https://www.bbc.com/news/world-asia-43660466 (accessed on 21 September 2018).
44. Dawn. Juniper Forest. Available online: https://www.dawn.com/news/1503085 (accessed on 21 September 2020).
45. Sarangzai, A.M.; Ahmed, A.; Siddiqui, M.F.; Laghari, S.K.; Akbar, M.; Hussain, A. Ecological status and regeneration patterns of *Juniperus excelsa* forests in north-eastern Balochistan. *FUUAST J. Biol.* **2013**, *3*, 53.

Article

Forest Cover Mapping Based on a Combination of Aerial Images and Sentinel-2 Satellite Data Compared to National Forest Inventory Data

Selina Ganz *, Petra Adler and Gerald Kändler

Forstliche Versuchs-und Forschungsanstalt Baden-Württemberg (FVA), 79100 Freiburg im Breisgau, Germany; Petra.Adler@Forst.bwl.de (P.A.); Gerald.Kaendler@Forst.bwl.de (G.K.)
* Correspondence: Selina.Ganz@Forst.bwl.de; Tel.: +49-761-4018-285

Received: 10 November 2020; Accepted: 10 December 2020; Published: 12 December 2020

Abstract: *Research Highlights*: This study developed the first remote sensing-based forest cover map of Baden-Württemberg, Germany, in a very high level of detail. *Background and Objectives*: As available global or pan-European forest maps have a low level of detail and the forest definition is not considered, administrative data are often oversimplified or out of date. Consequently, there is an important need for spatio-temporally explicit forest maps. The main objective of the present study was to generate a forest cover map of Baden-Württemberg, taking the German forest definition into account. Furthermore, we compared the results to NFI data; incongruences were categorized and quantified. *Materials and Methods*: We used a multisensory approach involving both aerial images and Sentinel-2 data. The applied methods are almost completely automated and therefore suitable for area-wide forest mapping. *Results*: According to our results, approximately 37.12% of the state is covered by forest, which agrees very well with the results of the NFI report (37.26% ± 0.44%). We showed that the forest cover map could be derived by aerial images and Sentinel-2 data including various data acquisition conditions and settings. Comparisons between the forest cover map and 34,429 NFI plots resulted in a spatial agreement of 95.21% overall. We identified four reasons for incongruences: (a) edge effects at forest borders (2.08%), (b) different forest definitions since NFI does not specify minimum tree height (2.04%), (c) land cover does not match land use (0.66%) and (d) errors in the forest cover layer (0.01%). *Conclusions*: The introduced approach is a valuable technique for mapping forest cover in a high level of detail. The developed forest cover map is frequently updated and thus can be used for monitoring purposes and for assisting a wide range of forest science, biodiversity or climate change-related studies.

Keywords: forest cover map; national forest inventory; aerial images; Sentinel-2; multisensory approach

---

1. Introduction

Forest cover is a key variable of interest to science, management and reporting [1]. Up-to-date and accurate information on the dynamics of forest cover is essential for the conservation and management of forests [2–4], carbon accounting efforts and the parameterization of a broad range of biogeochemical, hydrological, biodiversity and climate models [5]. A forest cover map provides information on the size, shape and spatial distribution of forests and thus assists in classifying the landscape into patterns. The spatial distribution of these landscape patterns is linked to the ecological functionality of the landscape [6] and provides new perspectives for ecological connectivity studies [7]. Therefore, the assessment of forest cover is aimed at facilitating decisions on biodiversity conservation and reforestation programs [8]. Furthermore, forest maps are crucial for global environmental change assessment and local forest management planning [7]. Especially in a changing climate, forest structures are changing very rapidly, meaning that knowledge of stocked areas and their changes is very important. Accordingly, forest information must be accurate, spatially detailed and up to date [9]. Due to the aforementioned reasons, there is an important need for a forest cover map containing a high level of detail.

Forests are inventoried for operational planning and forest management [9]. In many countries, National Forest Inventories (NFIs) are based on field surveys [10] and collect information on national forest resources. This enables strategic planning and policy development at the national level [9]. However, the aforementioned demand for information about forest cover exceeds the scope and design of most of the existing forest inventories [11]. Traditionally, forest cover maps are produced by visual interpretation of aerial images in combination with field surveys. Consequently, their development is time-consuming and therefore limited to relatively small areas [7].

In the last decade, various remote sensing-based methodologies have been used to map forest cover. The scales range from regional [12–23], nation-wide ([7] for Switzerland, [1] for Canada, [24] for Kyrgyzstan, [25] for Norway, [26] for Canada's Boreal Zone, [27] for Ukraine plus Eastern Europe and [3] for Japan), pan-European [28–30] and post-Soviet Central Asian [8] to the global level [31–36]. Besides aerial images and light detection and ranging (LiDAR) data, mainly satellite data have been used for the mapping and monitoring of forested areas and vegetation. Most studies involved either Sentinel-2 data [12,13,17,18,20,25,29] or LANDSAT data [1,26,27,30,31,36,37]. In contrast, the authors of [15] used SPOT data for forest cover and forest degradation mapping, whereas other studies used radar data like TanDEM-X [33], airborne S-Band radar [16] or Sentinel-1 [14]. The studies of [7,23], both based on digital surface models (DSMs) from image-based point clouds, produced wall-to-wall forest maps [7] and modeled fractional shrub/tree cover [23]. Furthermore, the work of [35] characterized global forest canopy cover using spaceborne LiDAR. In this context, the authors of [19] classified forest and non-forest based on full waveform LiDAR. Besides single sensor approaches using only one type of satellite or airborne data, multisensory approaches have been used for forest cover mapping. For example, [38] used LiDAR in combination with aerial images, [28] used SPOT 4/5, Corine Land Cover 2006 and LISS III, [3] used LANDSAT and Sentinel-1, [8] used LANDSAT together with MODIS data and [24] used a hybrid fusion strategy from Globe Land 30 2010 and USGS tree cover 2010. The authors of the aforementioned studies were mainly driven by the motivation to conduct change studies. Using time series analysis of LANDSAT images, the authors of [31] characterized global forest extent and change from 2000 to 2019. The studies of [12,13,20,21] each focused on the analysis of forest succession detection, height differences and forest cover dynamics as well as on forest conservation and deforestation. For validation, stand-level observations [25], maps of forest structure [26], high-resolution satellite data [15], LiDAR data [14], Bing or Google maps [3,37], Google Earth [18], orthoimages and terrestrial surveys [19] or NFI data [1,7,18] have been used.

Although there are numerous approaches in the literature to derive forest cover and forest cover changes, the derivation of forest areas and thus the differentiation between forest and non-forest is still a challenge. While available global or pan-European forest maps have a low level of detail and forest definition is not queried, local administrative data are, in many cases, oversimplified or out of date [7]. Another challenge is posed by the different definitions of the term "forest" at the regional, national and international levels. The term "forest" is a summarizing generic term which includes various forms of land cover, including tree-free forms, such as open spaces, wild meadows or traffic routes. In contrast, there are also tree-covered landscapes which are not defined or perceived as forests, including field shrubs, Christmas tree plantations or short-rotation plantations [39]. Land use is a key parameter in forest definitions but hardly assessable with remotely sensed data [7]. The most important quantitative characteristics of forest areas include minimum size, minimum width, tree height, minimum stocking level and stocking duration. A derivation of forest from remote sensing data according to the abovementioned definitions is not directly possible. However, it is technically feasible to record stocked and unstocked areas and to determine measurable criteria like height, area size and stand width [39].

The main objective of the present study was to generate a spatio-temporally explicit forest cover map, which takes the German forest definition into account. The key criteria are (1) minimum tree height, (2) minimum crown closure, (3) minimum width and (4) minimum area. The forest cover map based on the results of this study comprises the whole region of Baden-Württemberg, Germany, covering around 35,751.5 km^2. Furthermore, the forest cover map was compared to NFI data with respect to the presence or absence of forest. Finally, the differences between the forest cover map and NFI data were analyzed, categorized and quantified. To generate the forest cover map, we used a multisensory approach involving both aerial images and Sentinel-2 data. In fact, Sentinel-2 is designed for vegetation monitoring [40] and therefore multiple studies and projects exploited these data to generate vegetation maps. The spatial, temporal and radiometric resolution of Sentinel-2 enabled tree cover mapping with a medium level of detail; however, in combination with orthoimages from aerial image flights, a high level of detail could be achieved. With digital elevation models from image-based point clouds, vegetation height [39] was queried, resulting in the production of a tree cover and forest cover map. The introduced forest cover map does not include information about land use. In this context, biodiversity-relevant conditions or changes such as storm damages, permanent or temporary forest gaps or forested areas outside the forest area are mapped.

2. Materials and Methods

2.1. Study Area

The study area comprises the federal state Baden-Württemberg, located in southwest Germany. Baden-Württemberg is located between 7°30′ and 10°30′ E and 47°32′ and 49°47′ N. It covers 35,751.5 km^2, with an altitudinal range between 84 and 1493 m a.s.l. The state is characterized by different types of landscapes and is divided into seven main growing regions (see Figure 1).

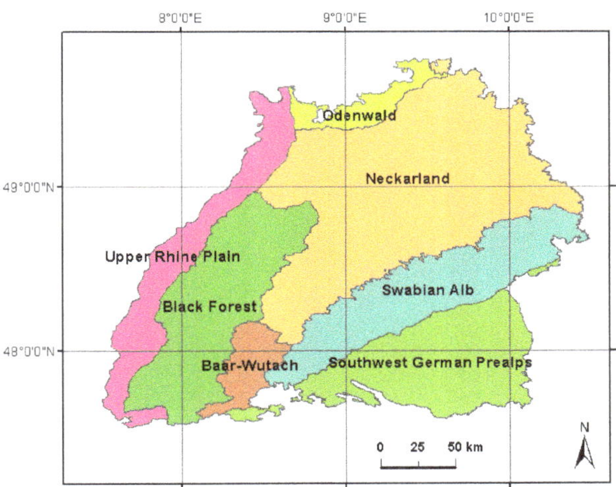

Figure 1. Main growing regions of Baden-Württemberg.

Baden-Württemberg has a varying regional proportion of forest cover with an average coverage of approximately 37.26% (±0.44%), which equals 13,319.6 km^2 (±157.8 km^2) in total. The forest ownership types are distributed as follows: corporate forests account for 40% of the forest area, private forests come to 35.9% and state-owned forests have 23.6%, whereas federal forests only cover 0.5%. Five tree species, namely, Norway spruce (*Picea abies* (L.) H. Karst), beech (*Fagus sylvatica* L.), silver fir (*Abies alba* Mill.), pine (*Pinus sylvestris* L.) and oak (*Quercus* sp. L.), occupy over 75% of the forest area and determine the forest character. Smaller percentages are found for ash (*Fraxinus excelsior* L.), Sycamore maple (*Acer pseudoplatanus* L.), Douglas fir (*Pseudotsuga menziesii* (Mirb.) Franco), larch (*Larix decidua* Mill.), hornbeam (*Carpinus betulus* L.), birch (*Betula* sp. L.) and alder (*Alnus* sp. Mill.). A total of 50 different tree species have been recorded in Baden-Württemberg, most of them with minimal proportions [41]. More detailed information on forests of Baden-Württemberg is provided in the third NFI report [41].

2.2. Remote Sensing Data

This section describes the used remotely sensed data. Aerial orthoimages and normalized digital surface models (nDSMs) allowed the extraction of vegetation and vegetation heights in a high level of detail, whereas Sentinel-2 data were used to derive vegetation in an additional, generalized approach.

2.2.1. Aerial Images

The orthoimages and nDSMs used in this study were processed from aerial images acquired by airborne image flights between 2011 and 2014. The aerial images provided by the state agency of spatial information and rural development of Baden-Württemberg (LGL) [42] contained the four bands red, blue, green and infrared (RGBI) at a radiometric resolution of 16 bits. Flight conditions such as time of day, season and weather conditions, as well as flight settings such as front/side overlap and camera type, as well as image-matching parameters such as the version of the image-matching software (SURE) and ground sampling distance (GSD), varied between the flight missions. The flights were conducted in the vegetation period, with the earliest flight on 15 April and the latest flight on 19 October. Front overlap was between 59% and 70%, whereas side overlap was between 28% and 40%. The aerial images were taken with UltraCam Eagle, UltraCam Xp, DMC 01 or DMC II. Photogrammetric point clouds and orthoimages were processed with a GSD of 40 or 50 cm using the software SURE of nFrames [43]. Detailed information on the settings of the image flights can be found in Table A1. As shown in Figure 2, the flight missions covered 99.7% of the study site.

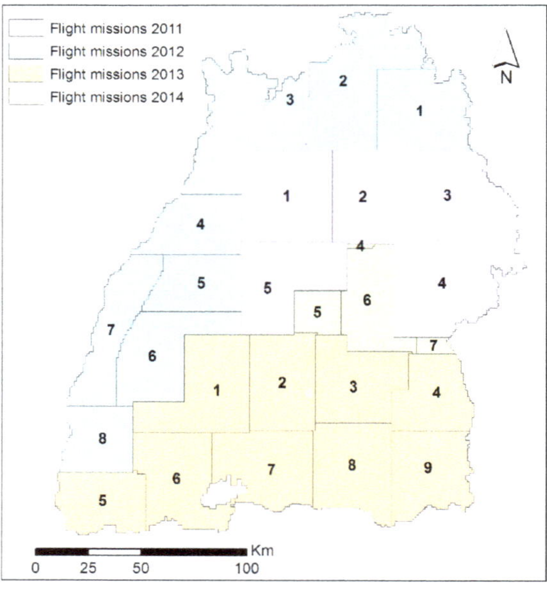

Figure 2. Flight missions in Baden-Württemberg between 2011 and 2014, showing parts of the flight missions used to derive the forest cover map.

2.2.2. Digital Surface Models

DSMs generated from photogrammetric point clouds served as the basis for deriving forest heights. Therefore, a combination of lasgrid and las2tin from LASTools [44] was applied. With lasgrid, the Z-value of the highest point per pixel was returned. Voids were filled with a square search radius of 3 pixels. Remaining voids were interpolated using the las2tin algorithm to triangulate the point cloud. The geometric resolution was set to 1 m. In order to obtain forest heights, the difference between the DSM value and the corresponding terrain height (=nDSM) was calculated. For this purpose, we used the LiDAR-based digital terrain model (DTM) of the LGL with a GSD of 1 m [45]. A detailed description about the applied methods for deriving aerial image-based nDSMs can be found in [46].

2.2.3. Sentinel-2 Data

In the frame of the Copernicus program, the European Space Agency (ESA) has launched the Sentinel-2 optical imaging mission. The Sentinel-2 mission involves a constellation of two polar orbiting satellites: Sentinel-2A and Sentinel-2B. Each satellite is equipped with the optical imaging sensor Multi-Spectral Instrument (MSI). The Sentinel-2 satellites provide images in the visible and infrared spectrum between 443 and 2190 nm and therefore are optimized for land surface observation [47]. Sentinel-2A was launched in 2015 and Sentinel-2B followed in 2017 [48]. According to [49], the resolution of up to 10 m and the scanning width of 290 km are ideal for detecting changes in vegetation and for making harvest forecasts, mapping forest stands or determining the growth of wild and agricultural plants.

In this study, we used Level-2A products of Sentinel-2. The processing of Level-2A products was conducted with Sen2Cor [48] and consisted mainly of scene classification and atmospheric correction. Scene classification is a necessary input for atmospheric correction. Atmospheric correction is performed in order to obtain bottom of atmosphere (BOA)-corrected transforms of the multispectral satellite imagery [50]. More information can be found in [48,50]. The image collections of Sentinel-2 (Level-2A) were compiled for the growing seasons (1 May to 31 September) of the years 2017, 2018 and 2019 taking all images into account with a cloud cover of < 25%. The time frame (2017 to 2019) was chosen due to the availability of Sentinel-2 (Level-2A) data, which have been available since 2017. The time gap between both NFI data and aerial images to the Sentinel-2 data was not expected to have much influence on the resulting forest cover map, as the vegetation layers were mainly used to remove buildings. The forest cover map is mainly determined by vegetation height derived via nDSM. To cover Baden-Württemberg, 9 Sentinel-2 granules were required. The used granules and the number of scenes per year are shown in Table 1. Since the date of recording and cloud cover varied between the Sentinel-2 granules, the number of the used scenes per granule and year differed.

Table 1. Number of Sentinel-2 granules used to calculate a vegetation mask.

Sentinel-2 Granules	2017	2018	2019	Σ
32UMU	3	4	6	13
32UMV	4	5	7	16
32UNA	8	8	9	25
32UNV	10	5	11	26
32UNU	5	1	7	13
32TMT	9	5	15	29
32TLT	14	19	36	69
32TNT	6	1	6	13
32ULU	7	19	34	60

2.3. NFI Data

The German NFI records a forest in its size and structure and thus provides information on the state of the forest and its development. In Germany, the first NFI was carried out in the years 1986 to

1988. The second NFI followed from 2001 to 2002 and the third NFI was from 2011 to 2012. Since 2010, a ten-year cycle has been in force for the NFI [41].

The present study used plot data of the latest NFI (2011–2012). The sampling design of the inventory utilizes a 2 × 2 km grid aligned on a systematic, national sampling grid. Grid points define the lower left corners of a square inventory tract of 150 × 150 m, whereas corner points of the tracts mark the centers of the NFI sampling plots [51]. Plots located in the forest were geolocalized with the Global Navigation Satellite System (GNSS) device MXbox of GEOSat GmbH [52] including a correction signal. At each tract corner point located in the forest, nested circular sub-plots of 5 different radii and a Bitterlich sub-plot [53], so-called angle count sampling (ACS), were established to record the set of inventory variables regarding trees, stand structure and site characteristics. With ACS, sample trees are selected with a probability proportional to their basal area, meaning that the range of each plot differs [54].

With the help of administrative forest cover maps and aerial photographs, new, previously non-forest tract corners in forests are identified ("preclarification"). Whether a tract corner is declared as being inside or outside of the forest is finally decided in the field ("forest decision", see Figure 3). The coordinates of the visited forest plots were corrected after the best attempt at reaching the point, whereas the non-forest plots have a tentative plot location. Out of the latest NFI, 35,744 NFI plots were available, while 13,299 were covered by forest, 425 were temporary or permanent unstocked areas and 22,020 were outside of the forest areas. Amongst other variables, tree heights were measured for a subset of sample trees. Subsequently, the tree heights of the remaining sample trees were modeled. Furthermore, information about land use, forest edges and accessibility was recorded.

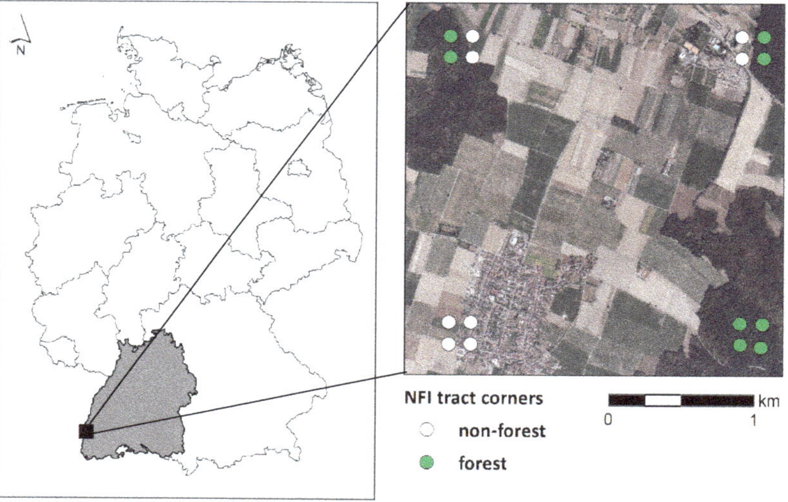

Figure 3. In Baden-Württemberg, the NFI tracts are aligned on a systematic, national sampling grid of 2 × 2 km. For each NFI tract corner, the presence/absence of forest is recorded ("forest decision").

To analyze forest cover within an NFI plot, the NFI tract corners were buffered with a radius of 10 m. This corresponds to the median distance between the NFI plot center and the maximum measured distance of all corresponding sample trees. The authors of [54] considered this approach as reasonable. In a further step, the percentage of forest cover within each plot was calculated. Since forest cover layer refers to land cover and NFI data to land use, the NFI data were filtered based on the available NFI plot information. For this purpose, only those NFI plots which were located in a forest with trees higher than 5 m or outside the forest area were evaluated. As a consequence, the following NFI plots were removed subsequently:

1. NFI plots at non-wood areas ($n = 425$);
2. NFI plots where the forest cover map could not be calculated due to clouds or artefacts ($n = 153$);
3. NFI plots which were not accessible ($n = 78$) and therefore no forest inventory was conducted;
4. NFI plots with maximum tree heights of < 5 m ($n = 381$);
5. NFI plots with stock per hectare $= 0$ (only calculated for trees with a diameter at breast height (dbh) of > 7 cm; $n = 278$).

In total, 1315 NFI plots were removed, equivalent to 3.68% of all plots. Subsequently, the NFI forest decision was compared to the forest cover map including 34,429 NFI plots (12,491 inside and 21,938 outside of the forest area).

2.4. Workflow

To calculate the forest cover, orthoimages and nDSMs as well as Sentinel-2 time series were used (see Figure 4). The challenge was to develop a method which could be flexibly adapted to the various aerial photographs and Sentinel-2 granules. Depending on the aerial image or satellite imagery, the spectral composition and lighting conditions varied.

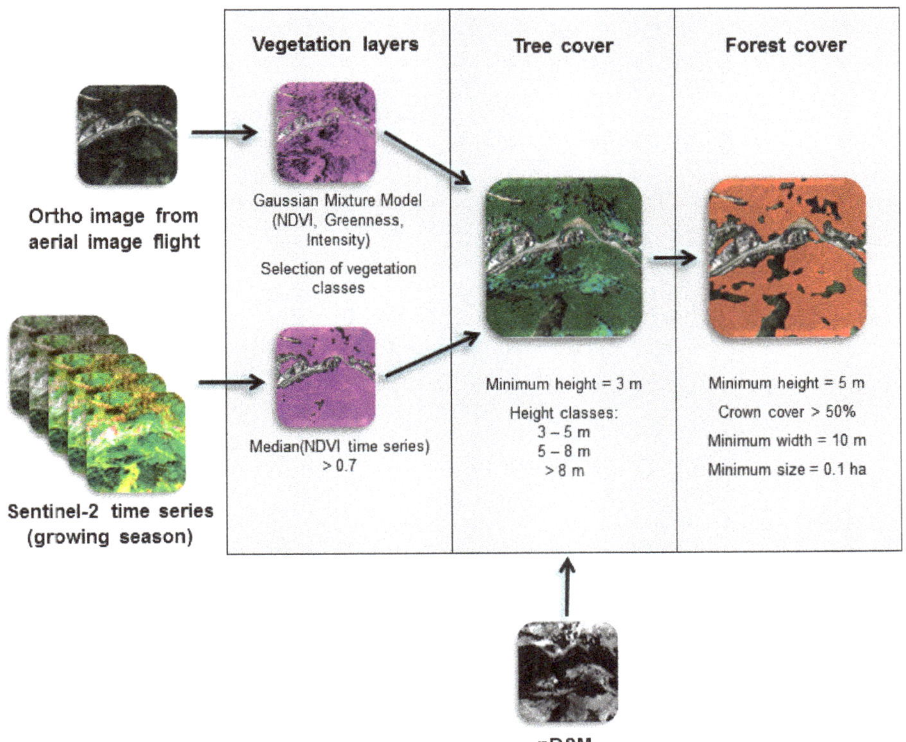

Figure 4. Calculation of the forest cover map. From both orthoimages and Sentinel-2 time series, a vegetation layer was built. By combining vegetation layers with nDSM, the tree cover layer was created. Forest cover is defined by the German NFI forest definition.

To derive a vegetation mask from the orthoimages, 5000 pixels inside and 5000 pixels outside the forest area were selected from 200 randomly chosen orthoimages for each mission. Administrative data about forest ownership types served as a forest mask. Based on the specific reflective properties of vital vegetation, parameters can be derived that serve to differentiate between inhabited surfaces and living or dead vegetation. The best known of these parameters is the normalized difference vegetation index (NDVI), which is calculated from the ratio of the values of the red (R) and near infrared (NIR) spectral range [39]. Besides NDVI, greenness and brightness were calculated for each pixel (see Formulae (1)–(3)).

$$NDVI = \frac{NIR - Red}{NIR + Red} \tag{1}$$

$$Greenness = \frac{Green}{Blue + Red} \tag{2}$$

$$Brightness = \sqrt{Blue^2 + Green^2 + Red^2} \tag{3}$$

Since the absolute values for these parameters differed greatly between the flight missions, we used a Gaussian mixture model (GMM) to select the parameters corresponding to vegetation. The parametrization of a GMM is based on the expectation–maximization (EM) algorithm for fitting the Gaussian distributions [55]. The EM algorithm iteratively approximates the ideal model of Gaussian distributions underlying the observed data. For initializing the EM approach, we assumed five components as land cover classes to be present inside the forest areas of Baden-Württemberg: coniferous trees, broad-leaved trees, vegetated non-wood areas, non-wood areas without vegetation (i.e., bare soil, bedrock, paved forest roads) and shadow. The basic parameters of the GMM were the number of components (n = 5), covariance type ("full") as the correlation of the used indices was in an acceptable range, maximum number of iterations (n = 100) and random state (n = 3) for reproducible results. The results of the GMM were 5 classes, which differed in terms of their value range of NDVI, greenness and brightness. The classes corresponding to vegetation were characterized by a high NDVI and greenness and were selected accordingly. The selection was visually verified by means of the orthoimages.

The second vegetation mask was derived from Sentinel-2 data. In order to eliminate all clouds and to obtain a homogeneous spectral composition for all granules, a time series was applied. From all Sentinel-2 scenes, the NDVI and subsequently the median NDVI of all scenes were calculated. The vegetation layer was created by setting a threshold value. This was possible due to the time series that achieved consistent NDVI values across Baden-Württemberg. By defining a threshold value (>0.7) for vegetation by visual interpretation of the orthoimages, a generalized vegetation layer of Baden-Württemberg was created covering forests along with shadow areas as well as meadows and green agricultural fields, without including settlement areas such as paved areas or rooftops. To generate a tree layer, the two vegetation layers were combined and compared with heights from the nDSM. Consequently, the tree layer consists of pixels, which were classified as vegetation based on orthoimages or Sentinel-2 data and have a minimum height of 3, 5 and 8 m determined by the nDSM. The forest cover map was calculated on the basis of this tree layer according to the forest definition by the German NFI, whereby the criteria canopy closure (>50%), minimum width (10 m), minimum size (0.1 ha) and minimum height (5 m) were considered.

The median of the NDVI time series of Sentinel-2 (Level2A) data was processed using the Google Earth Engine (GEE). The GEE computing platform offers a catalog of satellite imagery, geospatial datasets and planetary-scale analysis capabilities [56]. All further calculations were performed using a set of Python libraries for raster processing and analysis, GIS and scientific computing.

2.5. Comparison of Forest Cover Map and NFI Data

The forest cover map was compared to the forest decision (presence/absence of forest) of the NFI plot data (see Figure 5). To meet this objective, the confusion matrix, Kappa value and overall accuracy were calculated. NFI plot data served as reference data, whereas the forest cover map was used as a prediction. The threshold for forest presence was set to 50% forest cover, meaning that NFI plots with a forest cover of > 50% were classified as forest. Although the NFI plot data were filtered for this purpose, the remaining NFI plots do not completely correspond to forest cover. In the German NFI, minimum tree height is not defined, whereas it was an essential parameter for the derivation of the forest cover map. Furthermore, NFI data show land use, while the forest cover map represents land cover.

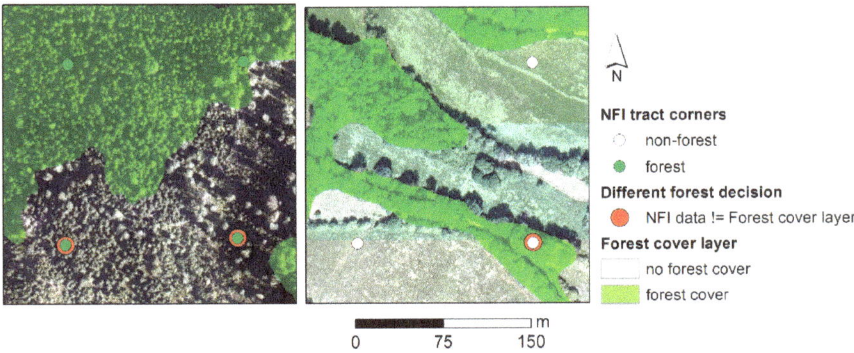

Figure 5. Comparison of NFI forest decision with forest cover map. Differences are outlined in red.

Based on the confusion matrix, differences between the forest cover map and NFI forest decision were identified and quantified. Therefore, the following data were used:

- Median nDSM heights within NFI plots: identification of forest plots < 5 m;
- Forest cover (%) of each plot: identification of plots at forest borders;
- Land use from the administrative layer: information about land use for each plot;
- Orthoimages: by visual interpretation, reasons for differences between the two datasets were identified and categorized.

3. Results

3.1. Forest Cover Map

As a result of this study, a wall-to-wall forest cover map for 99.71% of Baden-Württemberg was developed. It covers a total forest area of 13,222.1 km^2 with a spatial resolution of 1 m (see Figure 6). For 103.2 km^2 of the area (0.29% of Baden-Württemberg), no information (NoData) about forest cover could be generated due to missing aerial images, as described in Section 2.2.1. The final outcome of this study represents the first forest cover map of Baden-Württemberg in this very high level of detail. The distribution of forest cover appears to be reasonable. In particular, the high percentages of forest in the Odenwald, Black Forest and Swabian Alps as well as non-forested valleys in the Black Forest are clearly visible (see Figure 6). According to the forest cover map, approximately 13,222.1 km^2 (37.12%) of Baden-Württemberg is covered by forest. This is highly consistent with the results of the latest NFI [41], which reported a forest cover of approximately 13,319.6 km^2 (±157.8 km^2), equal to 37.26% (±0.44%).

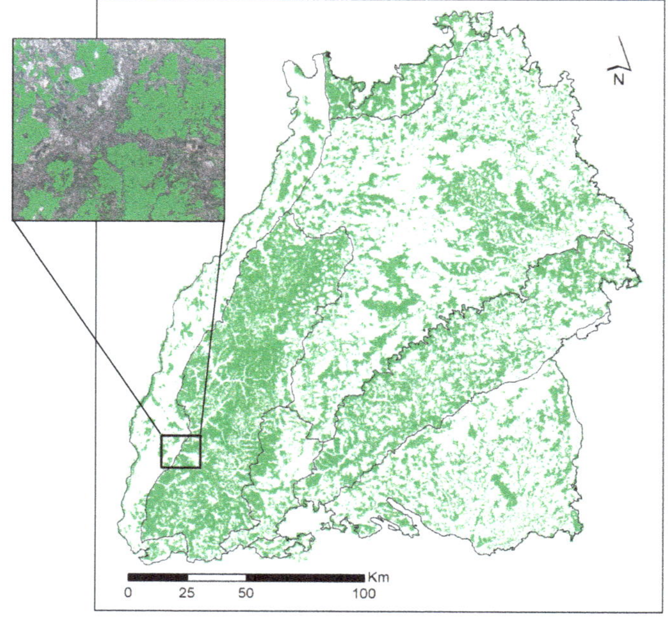

Figure 6. Forest cover map of Baden-Württemberg. The detailed map illustrates the forest cover in and around the city of Freiburg.

3.2. Comparison of NFI Plot Data and Forest Cover Map

The spatial agreement between the NFI data (based on a visual interpretation of forest presence/absence) and the forest cover map (based on remote sensing data) was 95.21% overall. The Kappa value showed an almost perfect agreement with a value of 0.90. The confusion matrix is shown in Table 2. The producer's and user's accuracies for both forest and non-forest were very similar. Non-forest achieved, with 96.06–96.41%, a slightly higher producer's and user's accuracy than the forest class with a producer's and user's accuracy of 93.12–93.73%.

Table 2. Confusion matrix of NFI plot data and forest cover map (fcm). OA = overall accuracy, UA = user's accuracy, PA = producer's accuracy, κ = Kappa value.

		NFI Plot Data		
		Non-Forest	**Forest**	**UA (%)**
	non-forest	21073	783	96.41
fcm	forest	865	11708	93.12
	PA [%]	96.06	93.73	
	OA [%]			95.21
	κ			0.90

3.3. Analysis of the Differences between NFI Plot Data and Forest Cover Map

The forest cover map based on remote sensing and the NFI forest decision differed by 4.79%, which is 1648 NFI plots in total. Each NFI plot, where the NFI forest decision was not corresponding with the forest cover layer (= "incongruent NFI plot"), was analyzed on the basis of median nDSM height, forest cover percentage and land use data. In combination with visual interpretation, the following reasons for the incongruence between the two datasets could be categorized (see Figure 7):

- Edge effects: Incongruences occurring at forest borders. Plots at forest borders were determined to have a forest cover of 1–99%.
- Forest definition: The NFI forest definition does not include a minimum tree height. For this reason, during the NFI, all stocked areas were classified as forest, regardless of tree height. In contrast, the forest cover layer defines areas with a height of < 5 m as non-stocked.
- Land use: Actual land cover does not correspond to land use. There can be non-stocked forest areas (e.g., storm damages) or stocked non-forest areas (e.g., orchards).
- Layer errors: differences due to errors in the remote sensing-based forest cover layer (e.g., errors in image matching lead to wrong nDSM heights).

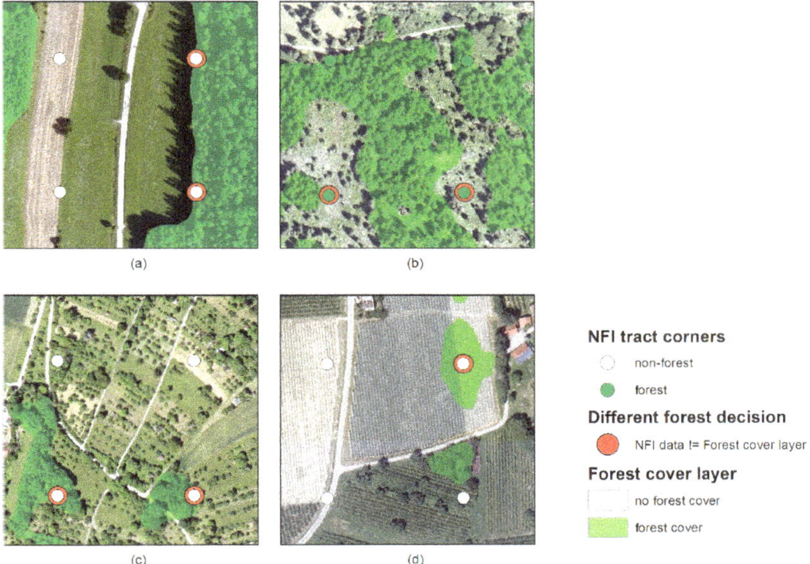

Figure 7. Categories of reasons for different forest decisions in the NFI plot data and forest cover map: (**a**) edge effects, (**b**) different forest definition (NFI does not define minimum height of trees), (**c**) land use not equal to land cover, (**d**) errors in forest cover layer. Differences are outlined in red.

Based on the defined categories, each incongruent NFI plot was assigned to a specific category. The distribution of these NFI plots is illustrated in Figure 8. It can be observed that edge effects are evenly distributed all across the study site. The category forest definition occurs most often in the northern Black Forest which is characterized, partly, by a high proportion of openings in forests. In contrast to the category forest definition, the category land use is slightly stronger when represented outside of forest-dominated regions. Only three errors in the forest cover layer were identified.

Statistically, the categories edge effect and forest definition are the most common, with 2.08% and 2.04% of all NFI plots, respectively. The category land use represents 0.66% overall, whereas layer errors occur only in three cases, yielding a presence of 0.01%. To analyze the proportion of forest presence/absence within the categories, the NFI forest decision served as a reference. Forest presence/absence is distributed unevenly within the categories. Edge effects occurred in 1.85% of the NFI-based non-forest plots and only in 0.23% of the forest plots, whereas all NFI plots categorized to the class forest definition were situated in the forest. From the category land use, 0.65% of the NFI plots were located outside the forest. All errors in the forest cover map (n = 3) occurred outside the forest areas (see Figure 9). The NFI plots categorized as errors in the forest cover map are shown in Figure 10. Two of them are lying in agricultural fields, whereas one plot is located on a rooftop of a building.

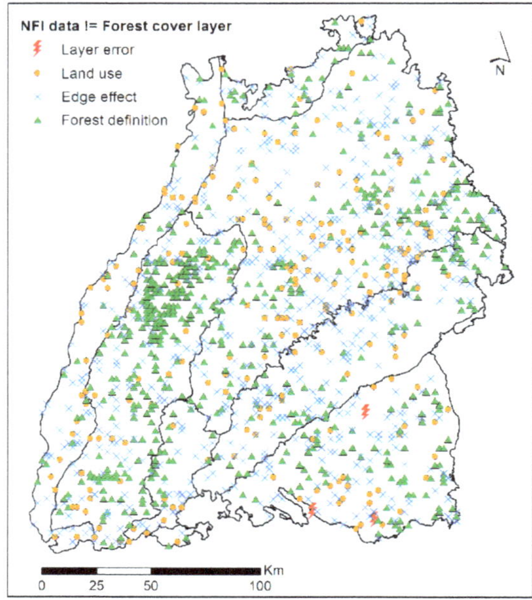

Figure 8. Spatial distribution of the incongruent NFI plots.

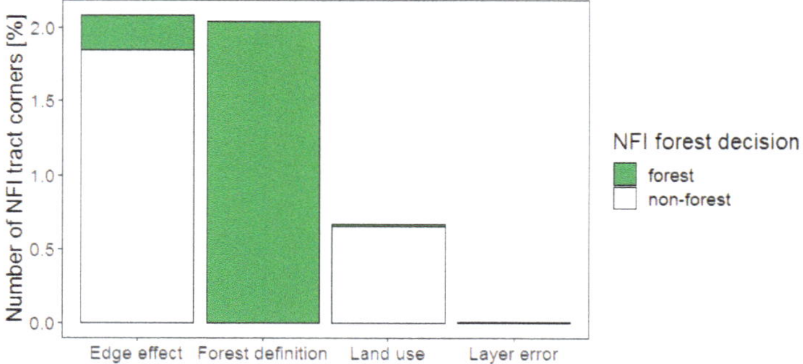

Figure 9. Statistical distribution of the incongruent NFI plots. The number of NFI tract corners is related to the total number of NFI tract corners evaluated.

Figure 10. NFI plots with different forest decisions in the NFI plot data and forest cover layer caused by errors in the forest cover layer. (**a**,**b**) Differences in agricultural fields, and (c) NFI plot located on arcoftop. Differences are outlined in red.

4. Discussion

4.1. Forest Cover Map

In the present study, the first wall-to-wall forest map of Baden-Württemberg was produced. We agree with the statement of the authors of [1], who argued that the remotely sensed forest area mapping within the inventory cycles of an NFI provides information that extends the sample-based NFI information content and improves the reporting capacity. The derivation of the forest cover map is almost completely automated, meaning that the applied methods are suitable for area-wide forest mapping. Only for the creation of the vegetation layer is a selection of GMM classes representing vegetation necessary. Sentinel-2-derived time series together with aerial images enabled spatially and temporally explicit depictions of the forest area. By picking the median NDVI of the Sentinel-2 time series, clouds as well as cloud shadows could be removed. Variations in illumination as well as differences in the reflection of vegetation due to different acquisition times were minimized. The multisensory approach combined the advantages of both sensors: Sentinel-2 data performed better inside of forests due to less shadow effects compared to aerial images; orthoimages complemented the Sentinel-2 data by their high resolution. Thereby, single trees outside the forest area and very small forest patches could be detected. Similar to [7], the spatial resolution of the forest cover map is 1 m. As described in [30], spatially detailed data are preferred due to the following reasons: First, applying coarse-resolution data results in a local overestimation of the dominant land cover class and consequently in lower fragmentation rates [57]. For example, road corridors as an important contributor to forest fragmentation [58] can only be detected if the images have a sufficiently fine spatial resolution. Second, due to the average size of forest patches in Europe, an accurate analysis of forest patches requires data with a spatial resolution finer than 100 m [30]. Aerial images are affected by changing illumination conditions both between and within images [54]. This affects the reliability and robustness of spectral metrics derived from aerial images. We managed to compensate the effect of changing illumination as well as the influence of the different settings between the flight missions (see Appendix A) by introducing the GMM and selecting the resulting GMM classes corresponding to vegetation. With the developed methodology, it was possible to map the forest cover in a consistently high quality over the entire area of Baden-Württemberg. The results of this study show that a forest cover map can be derived by aerial images acquired in numerous photogrammetric image flights with varying times of flight and camera types as well as different flight and camera settings. Across the forest cover map, errors are equally distributed, indicating that they are unaffected by the differences between the aerial image flights or the Sentinel-2 granules.

As the methodology is entirely based on the analysis of the spectral properties of land cover, the resulting map represents a forest cover map rather than a forest (land use) map. Furthermore, the combination of spectral and height information basically only captures the visible canopy. Objects below cannot be considered (e.g., parking lots or camping sites in the forest). Furthermore, the forest cover map contains also, for example, field shrubs, hedges, parks, orchards, short-rotation plantations and energy wood areas. On the other hand, temporary or permanently unstocked areas (e.g., timber stockyards, forest roads), which by definition are forest, are not recorded. Depending on the applicable forest definition and task, these areas must be added or subtracted [39].

4.2. Comparison NFI Plot Data and Forest Cover Map

With a few exceptions, the present forest cover map agrees well with the NFI data. The high overall accuracy of 95.21% of the forest cover layer indicates that the applied methods and results are promising. However, it should be considered that this kind of accuracy assessment takes into account all plots including plots at forest edges, plots with a canopy height of <5 m and plots where land cover does not match the actual land use. Since the forest cover layer only has the claim to cover forest without considering land use, the accuracy of the forest cover layer is much higher than the calculated overall accuracy. This is also evident in the small number of incongruences caused by layer errors. As mentioned in [7], a comparison with other studies is difficult. Studies, e.g., mapped forest with different input data types, had substantially smaller study areas or used varying forest definitions. To compare the results with other studies on forest cover, the studies mentioned in the Introduction can be reconsidered. In the following, a brief comparison is drawn with studies that also used NFI data as a reference: Ref. [7] obtained, with DSMs from image-based point clouds, an overall accuracy of 97% for the forest cover map of Switzerland taking the land use criterion into account; Ref. [18] used Sentinel-2 data for tree cover mapping for two study sites in Europe and achieved an overall accuracy of up to 90%; Ref. [1] used Landsat data to map the forests of Canada and achieved an overall spatial agreement of 84%.

Mapping of forest areas is affected by the categorical, spatial and temporal elements involved [1]. Different forest definitions as well as data processing and analysis methods lead to different mapping results [59]. In our study, we identified four reasons for incongruences between the NFI forest decision and forest cover layer: (a) edge effects, (b) different forest definitions, (c) land cover does not match land use and (d) layer errors.

The authors of [7] identified forest borders as one of the main reasons for mismatches between the remotly sensed forest cover map and NFI data. They revealed that in densely forested or entirely non-forested areas, the forest cover map has a higher accuracy than at forest borders. Similar to the aforementioned study, forest borders explained 2.08% of all incongruences. If an NFI tract corner was placed onto a road, no inventory was performed. Therefore, the category edge effect includes not only NFI plots on forest borders but also NFI plots on roads intersecting forests. Furthermore, the co-registration of field data with remote sensing data causes information uncertainties due to position inaccuracies. The NFI plots have an unknown positioning error due to inaccuracies of the GNSS signals below the forest canopy. This can lead to a shift between the observations in the field and in the remote sensing data [46]. Consequently, the category edge effects may also contain NFI plots with position errors.

In addition, the NFI does not define minimum tree height, which led to incongruences between the NFI forest decision and forest cover map. Forest definition is basically a subcategory of the category land use. The categories might overlap, since the assignment of non-stocked areas such as forest gaps to the category forest definition is based on the fact that land use is forest. If tree height is too low, but the plot is categorized by the NFI as forest, land use is forest, but land cover is not. That is the reason why these incongruent plots are all inside the forest. Based on the nDSM, a minimum vegetation height can be identified, which was applied in the form of a threshold value eliminating all objects that lie below this minimum height. When setting such a threshold value, it should be noted that the lower it is set, the more difficult it is to separate forest vegetation from other forms of vegetation. If the minimum height is set low, the likelihood of confusing forest vegetation with high-growing herbaceous vegetation, e.g., corn fields, large grasses and tall perennials, increases. Small trees in plantations and natural regeneration are not shown in the forest cover map due to their small size [39].

Generally, leisure facilities such as parks and sports fields, infrastructure such as roads (avenues) and cemeteries and agricultural areas such as green strips or orchard meadows are not considered as forests. Furthermore, in the category land use, there are 53 NFI plots where the forest decision is not verifiable. This might be explained by errors during preclarification. New, previous non-forest tracts, which are located with at least one corner in the forest, were overseen.

In addition, the defined minimum forest area of 0.1 ha takes also very small forests into account. Most of these areas are not determined as forest in terms of land use, but in the forest cover map, they are declared to be stocked.

The results show that there are almost no errors in the forest cover layer ($n = 3$). If the land surface depicted in aerial images is too homogeneous, image matching may be incorrect, resulting in artifacts in the orthoimages and wrong heights in the elevation models. Depending on the selected class of the Gaussian mixture model, it is possible that green, gray or light roofs in the orthoimages will be classified as vegetation. Most of the invalid areas are discarded as soon as the minimum height is considered using the nDSM. However, in the case of high-non-vegetation objects such as houses or high-voltage lines, non-vegetation may be included in the forest layer. Due to the Sentinel-2 pixel size of 10×10 m, parts of non-vegetation areas such as corners of rooftops can be included in the forest cover map.

The comparison of the remote sensing-based forest cover and the percentage forest cover of Baden-Württemberg reported by the latest NFI [41] shows a high correspondence with a slight difference of 0.14 percent points. Mismatches can be explained with the categories analyzed in Section 3.3. The fact that the reported forest area is higher than the calculated forest cover leads to the assumption that, in Baden-Württemberg, there are more forest areas with tree heights of <5 m than stocked areas outside of forests. This might be explained partially by the high proportion of openings in the northern Black Forest (see Figure 8).

5. Conclusions

Our study developed the first forest cover map of Baden-Württemberg using remote sensing data. The comparison with the NFI data showed that the applied methods and results are promising. Satellite-derived time series in combination with aerial images enable spatially and temporally explicit depictions of stocked areas in a high level of detail. Due to the free availability of the Sentinel-2 products as well as the regular image flights of the LGL, data availability is guaranteed on a long-term basis. The tree and forest cover maps are currently available for the years from 2011 to 2019 and will be updated regularly. In the near future, this will enable the elaboration of time series analysis and thus will allow long-term and area-wide monitoring of tree and forest cover over Baden-Württemberg. Not only in the future but already now, forest cover maps provide valuable information for multiple research questions in biodiversity studies, forest resource analysis, climate change research and other forest-related studies.

Author Contributions: Conceptualization and methodology, S.G. and P.A.; writing—original draft preparation, S.G.; writing—review and editing, S.G., P.A. and G.K.; supervision, project administration and funding acquisition, P.A. All authors have read and agreed to the published version of the manuscript.

Funding: This research was funded by "Sonderprogramm zur Stärkung der biologischen Vielfalt" (Land Baden-Württemberg).

Acknowledgments: We are grateful to Miguel Kohling for his contribution to method development and Python programming. Furthermore, we would like to thank Eva Wickert for her assistance in literature management and for proofreading the manuscript. We acknowledge the provision of the remote sensing and inventory data by the land surveying authority of Baden-Württemberg (LGL) and the forest service of Baden-Württemberg (ForstBW).

Conflicts of Interest: The authors declare no conflict of interest.

Appendix A

Table A1. Settings of the image flights.

Year	Mission	Date of Acquisition	Overlap [%]	Camera	SURE Version	GSD [m]
	1	12 July 2013	60/33	UltraCam Eagle	1.3.1.132	0.4
	2	13 July 2013 14 July 2013	62/37	UltraCam Eagle	1.3.1.336	0.5
	3	17 June 2013	64/38	UltraCam Xp	1.3.1.336	0.5
	4	19 June 2013	59/28	UltraCam Xp	2.2.1.1189	0.5
	5	17 June 2013	63/46	UltraCam Eagle	1.2.1.209	0.4
2013	6	18 June 2013 14 July 2013	62/40	UltraCam Xp	1.2.1.209	0.4
	7	13 July 2013 16 July 2013	64/38	UltraCam Xp	2.2.1.1189	0.5
	8	6 June 2013 7 June 2013 8 June 2013	63/34	UltraCam Eagle	2.2.1.1189	0.5
	9	19 June 2013 12 July 2013	59/31	UltraCam Xp	2.2.1.1189	0.5
	1	24 June 2012 30 June 2012	59/29	UltraCam Xp	3.0.0.0	0.5
	2	25 May 2012 23 July 2012	59/32	UltraCam Eagle	2.3.1.66	0.5
	3	23 July 2012	60/30	UltraCam Eagle	3.0.0.0	0.5
	4	15 April 2012	59/33	UltraCam Xp	2.3.1.66	0.5
2012	5	1 August 2012	61/40	DMC II	2.2.1.1212	0.5
	6	26 May 2012 25 July 2012 27 July 2012 1 August 2012 19 October 2012	61/39	DMC II	2.3.1.66	0.5
	7	23 July 2012	59/34	DMC 01	2.3.0.38	0.5
	8	25 July 2012	61/43	UltraCam Xp	1.3.1.336	0.4

Table A1. *Cont.*

Year	Mission	Date of Acquisition	Overlap [%]	Camera	SURE Version	GSD [m]
2011	1	5 May 2011 6 May 2011	66/34	DMC 01	3.0.0.0	0.5
	2	6 May 2011	62/32	UltraCam Xp	3.0.0.0	0.5
	3	6 September 2011 10 September 2011 16 September 2011	61/32	UltraCam Xp	3.0.0.0	0.5
	4	3 September 2011	62/31	UltraCam Xp	3.0.0.0	0.5
	5	5 May 2011	64/31	UltraCam Xp	2.3.0.38	0.5
2014	5	26 September 2014	61/34	UltraCam Xp	1.3.1.284	0.4
	6	22 June 2014 26 June 2014	70/45	UltraCam Eagle	1.3.1.284	0.5
	7	6 June 2014	60/31	UltraCam Xp	1.3.1.392	0.5

References

1. Wulder, M.A.; Hermosilla, T.; Stinson, G.; Gougeon, F.A.; White, J.C.; Hill, D.A.; Smiley, B.P. Satellite-based time series land cover and change information to map forest area consistent with national and international reporting requirements. *For. Int. J. For. Res.* **2020**, *93*, 331–343. [CrossRef]
2. Sexton, J.O.; Noojipady, P.; Song, X.-P.; Feng, M.; Song, D.-X.; Kim, D.-H.; Anand, A.; Huang, C.; Channan, S.; Pimm, S.L.; et al. Conservation policy and the measurement of forests. *Nat. Clim. Chang.* **2016**, *6*, 192–196. [CrossRef]
3. Sharma, R.C.; Hara, K.; Tateishi, R. Developing forest cover composites through a combination of Landsat-8 optical and Sentinel-1 SAR data for the visualization and extraction of forested areas. *J. Imaging* **2018**, *4*, 105. [CrossRef]
4. Stibig, H.-J.; Malingreau, J.-P. Forest cover of insular Southeast Asia mapped from recent satellite images of coarse spatial resolution. *Ambio* **2003**, *32*, 469–475. [CrossRef]
5. Hansen, M.C.; Stehman, S.V.; Potapov, P.V. Quantification of global gross forest cover loss. *Proc. Natl. Acad. Sci. USA* **2010**, *107*, 8650–8655. [CrossRef]
6. Ren, Y.; Wei, X.; Wang, D.; Luo, Y.; Song, X.; Wang, Y.; Yang, Y.; Hua, L. Linking landscape patterns with ecological functions: A case study examining the interaction between landscape heterogeneity and carbon stock of urban forests in Xiamen, China. *For. Ecol. Manag.* **2013**, *293*, 122–131. [CrossRef]
7. Waser, L.T.; Fischer, C.; Wang, Z.; Ginzler, C. Wall-to-wall forest mapping based on digital surface models from image-based point clouds and a NFI forest definition. *Forests* **2015**, *6*, 4510–4528. [CrossRef]
8. Yin, H.; Khamzina, A.; Pflugmacher, D.; Martius, C. Forest cover mapping in post-Soviet Central Asia using multi-resolution remote sensing imagery. *Sci. Rep.* **2017**, *7*, 1–11. [CrossRef]
9. White, J.C.; Coops, N.C.; Wulder, M.A.; Vastaranta, M.; Hilker, T.; Tompalski, P. Remote sensing technologies for enhancing forest inventories: A review. *Can. J. Remote Sens.* **2016**, *42*, 619–641. [CrossRef]
10. Tomppo, E.; Gschwanter, T.; Lawrence, M.; McRoberts, R.E. *National Forest Inventories. Pathways for Common Reporting*; Springer: Dordrecht, The Netherlands, 2010.
11. Alam, M.B.; Shahi, C.; Pulkki, R. Economic impact of enhanced forest inventory information and merchandizing yards in the forest product industry supply chain. *Socio-Econ. Plan. Sci.* **2014**, *48*, 189–197. [CrossRef]
12. Adjognon, G.S.; Rivera-Ballesteros, A.; van Soest, D. Satellite-based tree cover mapping for forest conservation in the drylands of Sub Saharan Africa (SSA): Application to Burkina Faso gazetted forests. *Dev. Eng.* **2019**, *4*, 100039. [CrossRef]
13. Barakat, A.; Khellouk, R.; El Jazouli, A.; Touhami, F.; Nadem, S. Monitoring of forest cover dynamics in eastern area of Béni-Mellal Province using ASTER and Sentinel-2A multispectral data. *Geol. Ecol. Landsc.* **2018**, *2*, 203–215. [CrossRef]
14. Dostálová, A.; Hollaus, M.; Milenković, M.; Wagner, W. Forest area derivation from Sentinel-1 data. *ISPRS Ann. Photogram. Remote Sens. Spat. Inf. Sci.* **2016**, *III-7*, 227–233.

15. Hojas-Gascon, L.; Belward, A.; Eva, H.; Ceccherini, G.; Hagolle, O.; Garcia, J.; Cerutti, P. Potential improvement for forest cover and forest degradation mapping with the forthcoming Sentinel-2 program. *ISPRS Int. Arch. Photogramm. Remote Sens. Spat. Inf. Sci.* **2015**, *XL-7/W3*, 417–423. [CrossRef]

16. Ningthoujam, R.K.; Tansey, K.; Balzter, H.; Morrison, K.; Johnson, S.; Gerard, F.; George, C.; Burbidge, G.; Doody, S.; Veck, N. Mapping forest cover and forest cover change with airborne S-band radar. *Remote Sens.* **2016**, *8*, 577. [CrossRef]

17. Norovsuren, B.; Renchin, T.; Tseveen, B.; Yangiv, A.; Altanchimeg, T. Estimation of forest coverage in northern region of Mongolia using Sentinel and Landsat data. *ISPRS Int. Arch. Photogramm. Remote Sens. Spat. Inf. Sci.* **2019**, *XLII-5/W3*, 71–76. [CrossRef]

18. Ottosen, T.-B.; Petch, G.; Hanson, M.; Skjøth, C.A. Tree cover mapping based on Sentinel-2 images demonstrate high thematic accuracy in Europe. *Int. J. Appl. Earth Obs. Geoinf.* **2020**, *84*, 101947. [CrossRef]

19. Straub, C.; Weinacker, H.; Koch, B. A fully automated procedure for delineation and classification of forest and non-forest vegetation based on full waveform laser scanner data. *ISPRS Int. Arch. Photogramm. Remote Sens. Spat. Inf. Sci.* **2008**, *XXXVII*, 1013–1019.

20. Szostak, M.; Hawryło, P.; Piela, D. Using of Sentinel-2 images for automation of the forest succession detection. *Eur. J. Remote Sens.* **2017**, *51*, 142–149. [CrossRef]

21. Tian, J.; Schneider, T.; Straub, C.; Kugler, F.; Reinartz, P. Exploring digital surface models from nine different sensors for forest monitoring and change detection. *Remote Sens.* **2017**, *9*, 287. [CrossRef]

22. Wang, Z.; Ginzler, C.; Waser, L.T. A novel method to assess short-term forest cover changes based on digital surface models from image-based point clouds. *For. Int. J. For. Res.* **2015**, *88*, 429–440. [CrossRef]

23. Waser, L.; Baltsavias, E.; Ecker, K.; Eisenbeiss, H.; Ginzler, C.; Küchler, M.; Thee, P.; Zhang, L. High-resolution digital surface models (DSMs) for modelling fractional shrub/tree cover in a mire environment. *Int. J. Remote Sens.* **2008**, *29*, 1261–1276. [CrossRef]

24. Jia, T.; Li, Y.; Shi, W.; Zhu, L. Deriving a forest cover map in Kyrgyzstan using a hybrid fusion strategy. *Remote Sens.* **2019**, *11*, 2325. [CrossRef]

25. Breidenbach, J.; Waser, L.T.; Debella-Gilo, M.; Schumacher, J.; Rahlf, J.; Hauglin, M.; Puliti, S.; Astrup, R. National mapping and estimation of forest area by dominant tree species using Sentinel-2 data. *arXiv* **2020**, arXiv:2004.07503. [CrossRef]

26. Matasci, G.; Hermosilla, T.; Wulder, M.A.; White, J.C.; Coops, N.C.; Hobart, G.W.; Zald, H.S. Large-area mapping of Canadian boreal forest cover, height, biomass and other structural attributes using Landsat composites and lidar plots. *Remote Sens. Environ.* **2018**, *209*, 90–106. [CrossRef]

27. Myroniuk, V.; Kutia, M.; Sarkissian, A.J.; Bilous, A.; Liu, S. Regional-scale forest mapping over fragmented landscapes using global forest products and Landsat time series classification. *Remote Sens.* **2020**, *12*, 187. [CrossRef]

28. Kempeneers, P.; Sedano, F.; Seebach, L.; Strobl, P.; San-Miguel-Ayanz, J. Data fusion of different spatial resolution remote sensing images applied to forest-type mapping. *IEEE Trans. Geosci. Remote Sens.* **2011**, *49*, 4977–4986. [CrossRef]

29. Langanke, T.; Büttner, G.; Dufourmont, H.; Iasillo, D.; Probeck, M.; Rosengren, M.; Sousa, A.; Strobl, P.; Weichselbaum, J. *GIO Land (GMES/Copernicus Initial Operations Land) High Resolution Layers (HRLs)–Summary of Product Specifications*; European Environment Agency: Copenhagen, Denmark, 2016.

30. Pekkarinen, A.; Reithmaier, L.; Strobl, P. Pan-European forest/non-forest mapping with Landsat ETM+ and CORINE Land Cover 2000 data. *ISPRS J. Photogramm. Remote Sens.* **2009**, *64*, 171–183. [CrossRef]

31. Hansen, M.C.; Potapov, P.V.; Moore, R.; Hancher, M.; Turubanova, S.A.; Tyukavina, A.; Thau, D.; Stehman, S.; Goetz, S.J.; Loveland, T.R.; et al. High-resolution global maps of 21st-century forest cover change. *Science* **2013**, *342*, 850–853. [CrossRef]

32. Lwin, K.K.; Ota, T.; Shimizu, K.; Mizoue, N. Assessing the importance of tree cover threshold for forest cover mapping derived from global forest cover in Myanmar. *Forests* **2019**, *10*, 1062. [CrossRef]

33. Martone, M.; Rizzoli, P.; Wecklich, C.; González, C.; Bueso-Bello, J.-L.; Valdo, P.; Schulze, D.; Zink, M.; Krieger, G.; Moreira, A. The global forest/non-forest map from TanDEM-X interferometric SAR data. *Remote Sens. Environ.* **2018**, *205*, 352–373. [CrossRef]

34. Shimada, M.; Itoh, T.; Motooka, T.; Watanabe, M.; Shiraishi, T.; Thapa, R.; Lucas, R. New global forest/non-forest maps from ALOS PALSAR data (2007–2010). *Remote Sens. Environ.* **2014**, *155*, 13–31. [CrossRef]

35. Tang, H.; Armston, J.; Hancock, S.; Marselis, S.; Goetz, S.; Dubayah, R. Characterizing global forest canopy cover distribution using spaceborne lidar. *Remote Sens. Environ.* **2019**, *231*, 111262. [CrossRef]

36. Zhang, X.; Long, T.; He, G.; Guo, Y.; Yin, R.; Zhang, Z.; Xiao, H.; Li, M.; Cheng, B. Rapid generation of global forest cover map using Landsat based on the forest ecological zones. *J. Appl. Remote Sens.* **2020**, *14*, 022211. [CrossRef]

37. Myroniuk, V.V. Forest cover mapping using Landsat-based seasonal composited mosaics. *Sci. Bull. UNFU* **2018**, *28*, 28–33. [CrossRef]

38. Immitzer, M.; Koukal, T.; Kanold, A.; Seitz, R.; Mansberger, R.; Atzberger, C. Abgrenzung der Natura 2000-Waldflächen. Klassifikation von Wald, Offenland und Latschenfeldern im bayerischen Hochgebirge unter Verwendung digitaler Luftbild- und Laserscannerdaten. *LWF Aktuell* **2012**, *88*, 49–51.

39. Ackermann, J.; Adler, P.; Aufreiter, C.; Bauerhansl, C.; Bucher, T.; Franz, S.; Engels, F.; Ginzler, C.; Hoffmann, K.; Jütte, K.; et al. Oberflächenmodelle aus Luftbildern für forstliche Anwendungen. Leitfaden AFL 2020. *WSL-Berichte* **2020**, *87*, 60.

40. Drusch, M.; Del Bello, U.; Carlier, S.; Colin, O.; Fernandez, V.; Gascon, F.; Hoersch, B.; Isola, C.; Laberinti, P.; Martimort, P. Sentinel-2: ESA's optical high-resolution mission for GMES operational services. *Remote Sens. Environ.* **2012**, *120*, 25–36. [CrossRef]

41. Kändler, G.; Cullmann, D. *Der Wald in Baden-Württemberg: Ausgewählte Ergebnisse der Dritten Bundeswaldinventur*; Forstliche Versuchs-und Forschungsanstalt Baden-Württemberg: Freiburg im Breisgau, Germany, 2014.

42. Landesamt für Geoinformation und Landentwicklung Baden-Württemberg (LGL). Geobasisdaten. 19 January 2018. Available online: www.lgl-bw.deaz.:2851.9 (accessed on 15 August 2019).

43. nFRAMES GmbH Stuttgart nFRAMES. Available online: https://www.nframes.com (accessed on 9 November 2020).

44. Rapidlasso GmbH Fast tools to Catch Reality. Available online: https://rapidlasso.com (accessed on 9 November 2020).

45. Landesamt für Geoinformation und Landentwicklung Baden-Württemberg (LGL). Digitale Geländemodelle (dgm). Available online: https://www.lgl-bw.de/unsere-themen/Produkte/Geodaten/Digitale-Gelaendemodelle (accessed on 9 October 2020).

46. Schumacher, J.; Rattay, M.; Kirchhöfer, M.; Adler, P.; Kändler, G. Combination of multi-temporal sentinel 2 images and aerial image based canopy height models for timber volume modelling. *Forests* **2019**, *10*, 746. [CrossRef]

47. Gascon, F. The Sentinel-2 Mission Products. In Proceedings of the First Sentinel-2 Preparatory Symposium, Frascati, Italy, 23–27 April 2012; ESA SP, Volume 707.

48. Main-Knorn, M.; Pflug, B.; Louis, J.; Debaecker, V.; Müller-Wilm, U.; Gascon, F. Sen2Cor for sentinel-2. In Proceedings of the Image and Signal Processing for Remote Sensing XXIII, Warsaw, Poland, 4 October 2017; SPIE, The International Society for Optics and Photonics: Bellingham, WA, USA, 2017; Volume 10427.

49. Deutsches Zentrum für Luft- und Raumfahrt e.V. Die Sentinel-Satellitenfamilie. Available online: https://www.d-copernicus.de/daten/satelliten/daten-sentinels (accessed on 22 October 2020).

50. Richter, R.; Louis, J.; Müller-Wilm, U. *Sentinel-2 MSI—Level 2A Products Algorithm Theoretical Basis Document*; S2PAD-ATBD-0001, Issue 2.0; Telespazio VEGA Deutschland GmbH: Darmstadt, Germany, 2012.

51. Riedel, T.; Hennig, P.; Kroiher, F.; Polley, H.; Schmitz, F.; Schwitzgebel, F. *Die Dritte Bundeswaldinventur (BWI 2012). Inventur-und Auswertemethoden*; Bundesministerium für Ernährung und Landwirtschaft: Bonn, Germany, 2017.

52. GEOSat MX Box. Available online: https://www.geosat.de/en/products/system-geometer-mx/geobox (accessed on 9 October 2020).

53. Bitterlich, W. Die Winkelzählmessung. *Allgemeine Forst- und Holzwirtschaftliche Zeitung* **1947**, *58*, 94–96.

54. Kirchhoefer, M.; Schumacher, J.; Adler, P.; Kändler, G. Considerations towards a novel approach for integrating angle-count sampling data in remote sensing based forest inventories. *Forests* **2017**, *8*, 239. [CrossRef]

55. Gaussian Mixture Models. Available online: https://scikit-learn.org/stable/modules/mixture.html (accessed on 8 December 2020).

56. Google Earth Engine. Available online: https://earthengine.google.com/ (accessed on 22 October 2020).

57. Riitters, K.; Wickham, J.; O'Neill, R.; Jones, B.; Smith, E. Global-scale patterns of forest fragmentation. *Conserv. Ecol.* **2000**, *4*, 3. [CrossRef]

58. Riitters, K.H.; Wickham, J.D. How far to the nearest road? *Front. Ecol. Environ.* **2003**, *1*, 125–129. [CrossRef]
59. Comber, A.; Carver, S.; Fritz, S.; McMorran, R.; Washtell, J.; Fisher, P. Different methods, different wilds: Evaluating alternative mappings of wildness using fuzzy MCE and Dempster-Shafer MCE. *Comput. Environ. Urban Syst.* **2010**, *34*, 142–152. [CrossRef]

Publisher's Note: MDPI stays neutral with regard to jurisdictional claims in published maps and institutional affiliations.

Review

Use of Remote Sensing Data to Improve the Efficiency of National Forest Inventories: A Case Study from the United States National Forest Inventory

Andrew J. Lister [1,*], Hans Andersen [2], Tracey Frescino [3], Demetrios Gatziolis [2], Sean Healey [3], Linda S. Heath [4], Greg C. Liknes [1], Ronald McRoberts [1], Gretchen G. Moisen [3], Mark Nelson [1], Rachel Riemann [1], Karen Schleeweis [3], Todd A. Schroeder [5], James Westfall [1] and B. Tyler Wilson [1]

[1] USDA Forest Service, Northern Research Station, York, PA 27410, USA; greg.liknes@usda.gov (G.C.L.); mcrob001@umn.edu (R.M.); mark.d.nelson@usda.gov (M.N.); rachel.riemann@usda.gov (R.R.); james.westfall@usda.gov (J.W.); barry.wilson@usda.gov (B.T.W.)
[2] USDA Forest Service, Pacific Northwest Research Station, Portland, OR 97205, USA; hans.andersen@usda.gov (H.A.); demetrios.gatziolis@usda.gov (D.G.)
[3] USDA Forest Service, Rocky Mountain Research Station, Ogden, UT 84401, USA; tracey.frescino@usda.gov (T.F.); sean.healey@usda.gov (S.H.); gretchen.g.moisen@usda.gov (G.G.M.); karen.schleeweis@usda.gov (K.S.)
[4] USDA Forest Service, Washington Office, Washington, DC 20250, USA; linda.heath@usda.gov
[5] USDA Forest Service, Southern Research Station, Knoxville, TN 37919, USA; todd.schroeder@usda.gov
*   Correspondence: andrew.lister@usda.gov

Received: 21 November 2020; Accepted: 14 December 2020; Published: 19 December 2020

Abstract: Globally, forests are a crucial natural resource, and their sound management is critical for human and ecosystem health and well-being. Efforts to manage forests depend upon reliable data on the status of and trends in forest resources. When these data come from well-designed natural resource monitoring (NRM) systems, decision makers can make science-informed decisions. National forest inventories (NFIs) are a cornerstone of NRM systems, but require capacity and skills to implement. Efficiencies can be gained by incorporating auxiliary information derived from remote sensing (RS) into ground-based forest inventories. However, it can be difficult for countries embarking on NFI development to choose among the various RS integration options, and to develop a harmonized vision of how NFI and RS data can work together to meet monitoring needs. The NFI of the United States, which has been conducted by the USDA Forest Service's (USFS) Forest Inventory and Analysis (FIA) program for nearly a century, uses RS technology extensively. Here we review the history of the use of RS in FIA, beginning with general background on NFI, FIA, and sampling statistics, followed by a description of the evolution of RS technology usage, beginning with paper aerial photography and ending with present day applications and future directions. The goal of this review is to offer FIA's experience with NFI-RS integration as a case study for other countries wishing to improve the efficiency of their NFI programs.

Keywords: national forest inventory; forest monitoring; remote sensing; forest sampling; inventory efficiency; Forest Inventory and Analysis

---

1. Introduction

1.1. Value of National Forest Inventory (NFI) Data: Management, Research, Policy Decisions

Many nations consider natural resource monitoring (NRM) systems to be critical sources of information for making decisions on natural resource management, planning, and policy. The value of NRM results lays in their usefulness for answering key monitoring or other science questions and making decisions. Without reliable information, management decisions can be ill-informed, leading to poor outcomes for ecosystem health and human well-being. The advancement of science also depends on NRM data; they can serve as the foundation for a diversity of both basic and applied science applications. To make investments in NRM cost-effective, it is vital that nations design highly-efficient, long-term NRM systems.

Forests cover 31%, or 4.06 billion ha, of the world's total land area [1]. An efficient National Forest Inventory (NFI) can therefore be a key component of a nation's NRM system. NFIs typically consist of a statistical sample-based system in which data on attributes relevant to stakeholders are collected on a set of NFI sampling units distributed within an area referred to as a population [2,3]. Key NFI attributes often include variables associated with trees (e.g., species, diameter, height, health status), wildlife habitat (e.g., vertical structure, standing dead trees), and the land on which the forest grows (e.g., land cover, land use, ownership, management type) [4]. It is common for NFIs to use observations based on remote sensing (RS) throughout the system, such as in data collection, estimation, and analysis and reporting [5–7].

Many countries' governments, such as that of the United States [8], mandate NFI information or otherwise provide NFI program direction. For example, recent legislation in the United States (the Agriculture Improvement Act of 2018 (P.L. 115-334, section 8632), commonly referred to as the 2018 Farm Bill), explicitly directs the USDA Forest Service (USFS) to find efficiencies in its NFI program through the use of advanced technologies such as RS and to engage other stakeholders in these efforts. The Forest Inventory and Analysis (FIA) program, which executes the NFI, has gained efficiencies and created new products for decades through careful investments in technologies like RS [9–11].

For the purposes of this review, we define RS data as those which are collected by instruments, typically mounted on an aerial or space-borne platform, that are not in direct contact with the subject of the observation. These data are often stored in the form of images or other spatially-referenced data types, and are typically produced when light or other radiation interacts with a target object and is received by a sensor mounted on the instrument. We discuss many types of RS data, including those from satellite, airborne and ground-based platforms employing passive optical sensors, as well as those from active sensors such as radio detection and ranging (radar) and light detection and ranging (lidar).

The goal of this review is to, using the FIA program's experience as a case study, provide examples of RS integration strategies for countries seeking to improve efficiency and add value to their inventories. It is not intended to be an exhaustive listing of RS activities in FIA; rather, it focuses on key work that is based on either FIA data or institutional knowledge generated by the program. Our aim is to offer FIA as a model for other NFIs by describing the statistical theory that supports efficiency gains with RS, the progression of FIA's use of RS data from its inception to modern times, and potential future directions. Our objective is to provide a better understanding of how RS technology has been and currently is integral to forest inventory science and forest management.

1.2. Value and Uses of FIA Data

The NFI of the United States is a well-documented case study of a multi-resource inventory that offers benefits to many types of stakeholders. Conducted by the USFS FIA Program, it has been in operation in various forms for nearly a century. It serves as the definitive source for forest resource information at the national and state scales [8,12,13]. The inventory consists of approximately 326,000 permanent, remeasured plots distributed across both forest and nonforest areas of the United States, its territories, and affiliated islands. Forest trees and certain land-type variables such as forest type and site index are only measured in portions of plots considered to be forest based on FIA's definition, and other land-type variables, such as land cover, use and, in some areas, ownership, are measured on all plots. In some urban and other areas, trees outside forest are measured on nonforest portions of plots, and on a subset of plots, additional forest health variables are measured. The base intensity is approximately one plot per 2400 ha in most areas, with plots assigned to random locations within hexagonal cells formed from a tessellation of the population area. Varying by portion of the country, between 10 and 20% of the plots are generally sampled each year with a spatially-balanced (panelized) design. Estimates are produced using post-stratified estimation, with RS-based maps used to create strata [14,15].

The inventory is foundational for annual and 5 year reports [16,17], Renewable Resource Planning Act Assessments [13,18,19], and the U.S. Report on Sustainable Forests [20,21]. In addition, the data collected by the FIA program are used for many scientific studies. In a review of literature published or in-press between 1976 and 2001, Rudis [22,23] found over 1400 research publications that relied at least in part on FIA data. Tinkham et al. [24] reviewed a subsample of more recent (1991–2017) research publications that used FIA data. Both studies reported that FIA data contributed to a large number of science applications, including the carbon cycle, forest products and growth, climate, forest health, biological diversity, and inventory statistics [24].

In addition to science applications, FIA data have been used extensively for both local and regional land management. Hoover et al. [25] reviewed several ways that they have been used by the USFS National Forest System (NFS) for forest management, including informing land management plan revisions, assessing progress toward management plan goals, monitoring wildlife habitat, and characterizing the status of key resources. In a similar study, Wurtzebach et al. [26] identified ways FIA data are used to satisfy NFS monitoring requirements, including providing information on structure, function, composition, watershed condition and trends, carbon stocks, insect and disease mortality, wildfire effects, at-risk species, timber suitability, and other required data elements. Dugan et al. [27] and Birdsey et al. [28] used FIA data and satellite imagery and its derivatives to model the impact of various types of forest disturbance on the carbon dynamics of NFS lands. Finally, Randolph et al. [29] and Vogt and Koch [30] identified both potential and demonstrated uses of FIA data for forest health and invasive pest studies, such as research on invasive plant species, risk assessment and mapping, and monitoring impacts of invasive pests.

Other important uses of FIA data abound, and include use by the forest products industry for economic projections and processing facility siting, by non-governmental organizations for supporting their advocacy activities, and by the general public for satisfying various information and educational needs. These examples are a subset of the large number of basic and applied science applications to which FIA data and expertise have contributed [31].

1.3. Background on Efficiency

In the following section, we introduce background needed to help explain how FIA has traditionally used RS data to improve efficiency. Survey sampling statistical concepts and statistical efficiency are first described. This is followed by a description of ways to consider how improving statistical efficiency through RS data integration improves economic efficiency of the NFI.

1.4. Improvement of Statistical Efficiency—Statistical Inference

NFIs generally subscribe to the basic principles of statistical inference whereby they express their estimates in probabilistic terms [32], primarily using confidence intervals. The confidence interval width is closely related to the precision of an estimate, i.e., the shorter the confidence interval, the greater the precision. Additional expressions of precision include variances and standard errors of estimates, margins of error, sampling errors, or simply uncertainty, all of which serve as metrics that a data user will consider when making judgments or taking actions based on FIA estimates. These precision metrics are, therefore, fundamental barometers of the usefulness of FIA data and estimates. Forestry professionals have developed a comfort level associated with using precision metrics when evaluating inventory results for making decisions, making their inclusion with FIA estimates critical. The primary factors that affect precision, confidence interval widths and, therefore, the efficiency of the estimation, are the sampling design, the sample size, the statistical estimator, and the mode of inference of which we consider two, design-based inference, and model-based inference [33].

1.5. Design-Based Inference

The key features of design-based inference are that each population unit is assumed to only one possible observation, and inferential validity derives from a probability sampling design as the source of randomization. Further, design-based estimators are generally unbiased or nearly unbiased, meaning that the mean of estimates over all possible samples obtained using the same sampling design and sample size equals the true value [33]. However, even with an unbiased estimator, the estimate for any particular sample may still deviate substantially from the true value. NFIs typically use one or more of three design-based statistical estimators: Simple expansion estimators, post-stratified estimators, and model-assisted estimators [34].

1.5.1. Simple Expansion Estimators

The simple expansion estimators are the most familiar, are used with simple random and systematic samples, and incorporate no RS or other auxiliary data. These estimators are often used to illustrate the relationship between inventory design and statistical efficiency via a fundamental precision metric, the variance of the mean. For a simple random sample, the simple expansion estimator of the mean is expressed as

$$\hat{\mu} = \frac{\sum_i^n y_i}{n} \tag{1}$$

where y_i is the observation of the attribute of interest on plot i, typically expressed on a per unit area (hectare) basis, and n is the sample size. The variance estimator for the estimate of the mean is expressed as

$$\hat{v}(\hat{\mu}) = \frac{s^2}{n} = \frac{\sum_i^n (y_i - \hat{\mu})^2}{n(n-1)} \tag{2}$$

where s^2 is the sample variance. The half width of the confidence interval (CI_{hw}) of the mean, which is based on the variance of the mean, is a commonly used precision index, and is defined as

$$CI_{hw} = t_{1-\frac{\alpha}{2}, n-1} \sqrt{\hat{v}(\hat{\mu})} \tag{3}$$

where $t_{1-\frac{\alpha}{2}, n-1}$ is the $100 \cdot (1 - \frac{\alpha}{2})$th percentile of the t distribution with $n-1$ degrees of freedom. Using survey sampling theory, a user evaluating an estimate of a mean obtained from an NFI in the context of its confidence interval knows that, had the survey been performed a very large number of times with identical methodology, $100 \cdot (1-\alpha)\%$ of the confidence intervals generated would contain the true, albeit unknown, value of the population mean [35]. For example, FIA's state reports typically report uncertainty in terms of relative standard errors, which are approximately equivalent to 68% CIs. Uncertainty reporting helps the user make informed decisions by taking into account uncertainty in the estimate caused by some combination of the sampling protocol (sample size or plot design characteristics) and variability of the attribute in the population.

1.5.2. Post-Stratified Estimators

If the intent is to increase the precision of estimates and, thereby, reduce confidence interval widths, additional information must be acquired. Multiple approaches for acquiring and using additional information are possible. As per Equation (2), variances can be decreased and precision can be increased by increasing the sample size, n. However, doing so incurs additional costs. Preferable options are to acquire the additional information with minimal additional cost. One option is to revise the plot configuration so that each plot captures more landscape information. This option typically entails increasing the plot area or using a cluster design wherein plots are divided into spatially separated subplots with the same aggregated area as a large single plot. However, depending on the spatial correlation of the attribute of interest, the additional information acquired may be less proportionally than the additional area inventoried.

Another option is to use auxiliary information in the form of strata that are related to the attribute of interest. Stratified random sampling can then be used to optimize the spatial allocation of plots relative to a criterion such as variance. The stratified estimator of the mean takes the form

$$\hat{\mu} = \sum_{h=1}^{H} w_h \hat{\mu}_h \tag{4}$$

where h indexes the strata, w_h is a weight proportional to the area of the stratum, and $\hat{\mu}_h$ is the within-stratum mean. The variance estimator is

$$\hat{v}(\hat{\mu}) = \sum_{h=1}^{H} w_h^2 \frac{\hat{\sigma}_h^2}{n_h} \tag{5}$$

where n_h is the within-stratum sample size and $\hat{\sigma}_h^2$ is the within-stratum variance of observations around the stratum mean.

Because most NFIs use at least a large proportion of permanent plots and have a long history of measurements for those same plots, they are reluctant to interrupt or lose those histories. As a result, NFIs seldom use stratified sampling designs because stratifications based on landscape or land cover features change over time and would require re-allocation of plots to strata with each new inventory cycle, thereby producing a loss of historical continuity for at least some plots.

However, for simple random and systematic sampling designs, a large proportion of the benefits of the additional information in strata can still be realized by assigning plots to strata independently of or subsequent to the sampling and then using the stratified estimator in Equation (4) and a variance estimator similar to Equation (5). This technique, characterized as post-stratification, has a long FIA history beginning with double sampling for post-stratification using RS data in the form of aerial photography as the source of stratification data in the 1980s [36], and more recently using satellite data as the source of stratification data [37,38]. FIA currently uses post-stratified estimation as its standard tool for calculating estimates of status of and trends in forest attributes [39]. The advantage of stratified approaches is that they use the additional auxiliary information to increase precision, but a disadvantage is that the rather coarse level at which the auxiliary information is aggregated does not fully exploit its potential.

1.5.3. Model-Assisted Estimators

The model-assisted regression estimators use the additional auxiliary information at the plot level, which represents a finer scale of aggregation than do strata. With these estimators, the attribute of interest is predicted for each plot using a parametric prediction technique such as linear or nonlinear regression, a non-parametric technique such as k-nearest neighbors, or a machine learning technique such as random forests. The model-assisted estimator of the mean has the form

$$\hat{\mu} = \frac{1}{N} \sum_{i=1}^{N} \hat{y}_i + \frac{1}{n} \sum_{i=1}^{n} (y_i - \hat{y}_i) \tag{6}$$

where N is the population size, n is the sample size, and $\hat{y}_i$ is the prediction for the ith plot. The unbiasedness or near unbiasedness of the model-assisted regression estimator derives from the second term of Equation (6), which compensates for prediction error. Model-assisted variance estimators are based on deviations between observations and their predictions and take the form

$$\hat{v}(\hat{\mu}) = \frac{1}{n(n-1)} \sum_{i=1}^{n} (\varepsilon_i - \bar{\varepsilon})^2 \tag{7}$$

where $\varepsilon_i = y_i - \hat{y}_i$ and $\bar{\varepsilon} = \frac{1}{n} \sum_{i=1}^{n} \varepsilon_i$. If the predictions are sufficiently accurate, deviations between observations and their predictions should be smaller than deviations between observations and strata means as is the case with the stratified and post-stratified estimators or between observations and overall means as is the case with the simple expansion estimators. The result is that model-assisted regression variances and associated confidence interval widths are often smaller than corresponding variance and confidence interval widths for stratified and simple expansion estimators.

McRoberts et al. [40–42] demonstrated the superiority of model-assisted inference and post-stratification for estimates of FIA attributes compared to methods not using RS observations. Magnussen et al. [43], in their own work and a review of several other similar studies, also provide strong support for the use of RS data in a model-assisted context as a way to improve statistical efficiency. The specific mathematical linkages between post-stratified and regression estimation can be seen in Bethlehem and Keller [44], Breidt and Opsomer [45], and Stehman [46]. In addition, McConville et al. [11] provide a tutorial on model-assisted inference for forest inventory that presents post-stratification as a special case of a generalized regression estimator.

1.6. Model-Based Inference

The assumptions underlying model-based inference differ considerably from those for design-based inference [47]. First, each population unit has an entire distribution of possible values for an attribute, unlike design-based inference for which there is only a single value. Randomization with model-based inference is realized in the particular value that is observed for each population unit, not via selection of the sample. Finally, the validity of model-based inference is not based on probability samples but rather on the model of the relationship between the attribute of interest and the auxiliary information. An important consequence is that model-based inference can use, but does not require, probability samples. Therefore, model-based inference can be used for small areas for which probability samples are too small for reliable design-based inference and for remote and inaccessible areas for which no probability samples are possible [48]. However, the price to be paid for this greater applicability is that the model-based estimator of the mean is not necessarily unbiased. McRoberts et al. [49] provide the model-based estimator of the mean and estimator of the corresponding variance.

Naturally, the field of model-based inference is very large, and models may take many different forms and operate at different scales of resolution. As an example, there is a growing need for FIA estimates and information over smaller geographic areas, for shorter time periods, and for specific thematic groups. Efforts are underway to expand FIA's capacity to produce estimates for these smaller domains where there are too few plots to use design-based estimators that rely only on data within the domains of interest. Small area estimation (SAE) has been studied extensively in the statistical literature [50]. It relies on indirect methods that borrow strength from outside the domains of interest, integrating both plots and auxiliary RS data through models. There have been numerous examples in U.S. forest inventory history of using SAE methods to essentially increase the effective sample size within small domains. In particular, Moisen et al. [51] applied empirical best linear unbiased predictors (EBLUPs) to estimate forest area and biomass within burned perimeters in the Interior West. Lemay and Temesgen [52] compared imputation methods for modeling basal area and stem density. Goerndt et al. [53] compared synthetic, composite, EBLUP, and most similar neighbor approaches for estimating tree density, diameter, basal area, height, and cubic stem volume using lidar in Oregon. Mauro et al. [54] compared unit and area-level EBLUPs for constructing small area estimates of stand density, volume, basal area, quadratic mean diameter, and height in a western coastal area. Goerndt et al. [55] applied composite estimators to estimate values of FIA attributes relevant to bioenergy production over parts of a 20-state region in the northern US. These are just a few examples, and the field is growing rapidly.

Advantages of model-based inference include its ability to produce defensible estimates for smaller domains and the lack of reliance on probability samples. Disadvantages include its dependence on the assumed model and the consequent potential for bias [33]. Furthermore, those familiar with estimates based on traditional survey samples might prefer them over less familiar uncertainty metrics.

1.7. Hybrid Inference

Hybrid inference combines features of both design-based and model-based inference [56,57]. Hybrid inference is used when probability samples are available, but data for the sample units are model predictions with non-negligible uncertainty rather than observations subject to at most negligible measurement error. Hybrid inference incorporates uncertainty from both sources: The model-based prediction uncertainty for the individual sample units and the design-based sampling variability [58,59]. Hybrid inference is more applicable than is generally recognized. For example, FIA plot-level volume data are generally considered "observations" when, in fact, they are really aggregations of tree-level allometric model predictions. Incorporating the effects of this allometric model prediction uncertainty into the uncertainty of large area estimates of mean volume requires techniques such as hybrid inference [58]. Saarela et al. [60] and Ståhl et al. [59] document the utility of hybrid inference for inventory problems for which the model predictions are based on RS data.

1.8. Improvement of Economic Efficiency

There are two ways to look at how improving statistical efficiency through RS data integration could improve economic efficiency of the FIA survey. The first is by allowing FIA to meet NFI precision requirements with fewer field plots. For example, the precision targets for the FIA attributes *total timberland area and total volume of timberland growing stock trees* in the Eastern United States are 3% sampling error per 404,686 ha of timberland and 5% sampling error per 28,316,847 m^3 of wood volume, respectively [61]. The FIA sample was designed with these targets in mind, but through integration of RS as described in the discussion surrounding Equations (1)–(7), targets could be met in principle with fewer plots. McRoberts and Tomppo [5] give an example of this principle from FIA; they found that in Minnesota, USA, integrating RS data through stratification led to meaningful cost reductions compared to what would have been achieved using a simple random sample. Brooks et al. [62] found that post-stratification for change and forest status variables increased precision by over 100% compared to estimation without using RS data. In the states under the purview of the USFS Rocky Mountain Research Station, it was estimated that there would be an average saving of $300,000 (USD) per state per inventory cycle by transitioning from acquisition and manual interpretation of paper photographs to using RS imagery to perform stratification [63]. The same types of improvements can be shown to occur as the correlation between the attribute of interest and the RS data source increase through improvements in technology. For example, Köhl et al. [64] found that in a study in which a simulated lidar covariate was used to construct regression models for aboveground biomass, more than 800% more plots would be required to achieve 10% sampling error using a model with a coefficient of determination (R^2) of 0.3 compared to using one with an R^2 of 0.9.

The number of plots needed to meet precision requirements depends on the variable of interest, with attributes with highly variable values requiring more plots. Therefore, for the same number of plots, different attributes will have different precision estimates. A decision to reduce the number of field plots collected is therefore complicated, and entails costs associated with administrative overhead, loss of historical continuity, and disappointment of stakeholders who are using the data for a multitude of purposes including calibrating or validating RS-based models. However, in the event of budget reductions, RS integration and resulting precision improvement may help offset the negative impacts of forced cuts in field work. Perhaps more importantly, RS can add value by exceeding the precision targets that were initially set for the key attributes, and improving precision estimates for all other attributes that, without RS integration, would not have been as useful for making decisions.

The second economic benefit of RS data integration is also related to adding value to the program, through investment in products that go beyond traditional inventory summaries and analyses. Cost savings from reductions in required sample size can be invested in the production of geospatial products such as wall-to-wall maps and models that are used in myriad ways, examples of which are described above [9]. Maps are now part of the currency used for making decisions—they not only provide a synoptic depiction of the resource being reported, but they also provide tools that can be used by others through Geographic Information System (GIS) analyses. It is the pursuit of these efficiencies (statistical and economic) that have led to many of FIA's technological advances in RS, and will lead to further innovation.

A third benefit of RS data integration comes in the form of avoiding unnecessary field work. For example, FIA already makes substantive use of aerial photographs to collect observations on nonforested locations, reducing substantial travel costs to places where no physical measurements are required. Furthermore, RS offers promise for measuring forest attribute data that lend themselves to remote observation, such as land use and land cover change. Examples of these types of processes are given in the following sections.

It should be noted that integrating RS data also can incur costs. If imagery needs to be purchased (as was the case with Landsat satellite imagery in the past), it is especially costly. New computer software and hardware, additional computing requirements, and capacity building may also be required to implement the RS integration. However, in modern times, data sharing policies often result in imagery at no cost to the user, and sufficiently-powerful and ubiquitous software and hardware make RS integration costs relatively small.

2. Progression of FIA's Use of RS Data: Inception to Modern Times

2.1. Early Use of RS Data in FIA

FIA and other inventory programs have used RS observations operationally for decades. As RS technology has evolved, technologies to take advantage of new, higher quality, and more voluminous data have emerged. In this section, we first discuss the use of photointerpretation from analog sketch maps to aerial photography. Digital imagery from a variety of satellites allowed for improvements; we focus the discussion in this section on research activities and operational products from AVHRR, Landsat, and MODIS.

2.2. Photointerpretation (PI)

The earliest maps of forests in the United States were manually drawn from a combination of field reconnaissance and a primitive form of ocular RS by early census takers and cartographers in the late 1800s [65]. After World War 2, however, aerial photography became widespread and its use in forestry expanded greatly [66]. Although it was used for field work planning and other logistics, its principle use in FIA was as the first phase of what is termed two phase or double sampling [34,67,68]. In this statistical method, a set of points located on a dense grid were interpreted and assigned an attribute class, such as a land use or tree volume class. These classes were aggregated into strata that were useful for allocating plots to the landscape in ways that were more economical than would occur without using the technique. Most of the state-level inventories associated with what was to become the FIA program were performed using variations of this principle [69].

A noteworthy, value-added analysis produced with aerial imagery was conducted in the Northeastern FIA region by Riemann and Tillman [70]. In this unique study, PI points were assigned values for forest fragmentation metrics, and the fragmentation status of several Northeastern states was characterized. In the late 1990's, high resolution digital aerial orthophotos from the USGS became widely available for download or ordering on compact discs (CDs) or digital versatile discs (DVDs) [71]; this innovation made FIA pre-field logistical work and research studies that rely on aerial imagery much easier by allowing for analyses such as double sampling or PI of fragmentation to be conducted in a GIS. In the early 2000's, the USDA National Agriculture Imagery Program (NAIP) began to produce nationwide, high resolution digital aerial imagery on a recurring basis, opening up many new research and applications opportunities for FIA [72].

2.3. AVHRR

The cost of PI was large; in order for FIA's double sampling estimation approach to work, hundreds of thousands of points needed to be interpreted on thousands of paper photographs by large teams of full time interpreters. Peterson et al. [63], Reams and Van Deusen [73], and Wynne et al. [74] discussed the benefits of transitioning away from aerial photography toward operational use of digital satellite imagery in forest inventory; these included cost improvements, avoidance of reliance on national aerial imagery programs, and the superior temporal, spatial and spectral coverage of satellite RS data. However, Teuber [75] identified a major challenge associated with the operational use of RS imagery over large areas like those covered by FIA: data volume. To mitigate this concern, Teuber [75] proposed large area mapping with 1 km pixel data from the Advanced Very High Resolution Radiometer (AVHRR), which was carried on National Oceanic and Atmospheric Administration (NOAA) polar-orbiting weather satellites [76,77]. Software and hardware tools to process AVHRR data had been extensively used during the 1980s to make regional maps of forest attributes [78], with the goal of operationalizing the procedures. Teuber [75] found that AVHRR-derived forest area estimates were similar to those from the FIA program for several southeastern states, a promising finding that gave momentum to efforts to move toward implementing satellite RS data usage in FIA.

Research began to show how FIA could operationalize the use of AVHRR. For example, Zhu [79] found that model-based estimates of forest area from AVHRR data compared well with traditional FIA-based estimates for parts of the southern U.S. Roesch et al. [80] successfully used model-assisted inference with AVHRR imagery and FIA data to update county-level estimates of forest area between inventory cycles in Alabama. Moisen and Edwards [81] found that using AVHRR-based maps for stratifying FIA plots to improve estimates was only marginally less efficient than using more costly methods. Several studies employing model-based mapping with FIA and AVHRR data ensued, most notably, that of Zhu and Evans [82], which combined FIA and AVHRR data to produce the first 1 km pixel, nation-wide map of forest type. Cooke [83] built upon the Zhu and Evans [82] approach to refine the AVHRR-based models for parts of Texas and Oklahoma, compared results to FIA estimates, and proposed a forest mapping system that could be operationalized for use in FIA.

2.4. Landsat

Data from the Landsat satellites became more heavily used by FIA during the 1980s and 1990s, as well. Landsat 4, carrying the Landsat Thematic Mapper (TM) sensor, was launched in 1982, offering the 30 m × 30 m pixel size spatial resolution that aligns more closely with phenomena of interest to foresters than did most alternatives [84]. There were two problems with operational use of Landsat in FIA: cost and processing capabilities. In terms of cost, The Landsat Remote Sensing Commercialization Act of 1984 led to the privatization of Landsat data, and thus costs became an impediment, as scenes could cost hundreds of dollars each. Lister [85] describes a typical workflow for going from raw Landsat scenes to a finished large area map of forest cover; this includes obtaining multiple scenes from the same time period to deal with cloud cover, and from different time periods to exploit phenological information. Even with government-wide cost sharing agreements, the cost to purchase this large quantity of scenes impeded widespread adoption of the technology beyond small study areas for many government users, including FIA [84].

None-the-less, several state- and region-scale projects using Landsat data were conducted by FIA or partners. Cooke and Hartsell [86], in preparation for transitioning to the use of Landsat data for FIA estimates of forest area in Georgia, explored the relationships between FIA plot geometry and spectral properties of the Landsat data. McRoberts et al. [37,87,88] described an approach to using classified Landsat imagery for improving FIA estimates through stratification. A series of tests based on assigning FIA plots, under a post-stratified design, to strata constructed with Landsat data were performed around the country by, for example, Hansen and Wendt [36] for the states of Indiana and Illinois, Hoppus et al. [89] for the state of Connecticut, Moisen et al. [90] for parts of FIA's Intermountain West region, Dunham et al. [91] in the Pacific Northwest, and Cooke and Jacobs [92] in Georgia. These tests showed that it was feasible to operationalize the use of Landsat and other RS data-based maps in a variety of forest ecosystems to improve precision via stratification.

Other mapping agencies saw the value of partnering with FIA to help calibrate and validate their Landsat-based maps, and FIA saw the benefits of leveraging these partnerships to obtain maps that met its needs. For example, the Multi-Resolution Land Characterization (MRLC) Consortium was tasked with producing a National Land Cover Database (NLCD) consisting of maps and other products that characterize the nation's land cover [93,94]. For the 2001 NLCD products, MRLC prepared Landsat and other data, and FIA provided plot data for both calibration and validation of maps [95]. While FIA scientists had previously demonstrated their ability to produce their own maps, it became evident that leveraging NLCD maps, and partnering with the MRLC so that FIA needs were more likely to be met, was more efficient economically. The knowledge base associated with using NLCD in FIA workflows grew to the point where NLCD maps were adopted as the foundational layer for FIA's use of post-stratification for most of the country during the first decades of the 2000's [37,69,89,91,96].

Other, later examples of how FIA partners combined FIA data with Landsat imagery for national scale mapping include the multi-agency Monitoring Trends in Burn Severity (MTBS) project [97] and the Landscape Fire and Resource Management Planning Tools (LANDFIRE) program [98,99]. The goal of MTBS is to map fire severity and extent of fires in the U.S. since 1984, using information derived from Landsat imagery. LANDFIRE, which completed its first national product in 2009 and uses MTBS data, produces national-scale geospatial data for use in fire planning, management, and operations. LANDFIRE is an invaluable resource to the fire community, and is heavily based on FIA data for calibration and validation [100,101]. FIA scientists have likewise used LANDFIRE and MTBS products in their research. For example, Whittier and Gray [102] and Shaw et al. [103] intersected FIA plot data with MTBS maps in order to characterize the effects of fire on forests. Dugan et al. [27] used FIA data, Landsat imagery, MTBS, and other data to characterize the disturbances associated with carbon loss on USFS NFS land. All of these applications provide evidence for the value of partnerships with agencies outside of FIA to both acquire expertise and gain access to data products that would otherwise be unavailable, and of the value of using FIA field plot data with RS data.

2.5. MODIS

The 1999 launch of the Moderate Resolution Imaging Spectroradiometer (MODIS) on NASA's Terra satellite spurred further innovation into RS integration with FIA. MODIS imagery had several advantages over Landsat data at that time: It was free, it had a near-daily revisit cycle, and it had large swath widths. However, its disadvantage was its spatial resolution—the pixel sizes in wavelengths relevant to vegetation monitoring are 250 and 500 m [104], which, though better than AVHRR, are still quite large. None-the-less, trials of synergistic uses of MODIS with FIA data were begun in the early 2000s. Initial work focused on comparisons of FIA estimates with those derived from MODIS products. For example, White et al. [105] conducted an accuracy assessment of a 500 m pixel MODIS tree cover product (Vegetation Continuous Fields, or VCF) using over a thousand FIA plots. This study, and that of Nelson et al. [106], which also compared FIA to VCF estimates of forest area, found a linear relationship between FIA-based estimates and those from classified VCF products, but that VCF overestimated tree cover for low values of FIA forest cover and underestimated for large values. This was likely due to the effects of spatial resolution on map-based estimates [107] or the spatial mismatch between FIA plots (which consist of triangular clusters of four 0.017 ha subplots distributed within a 44 m radius circular area) and the MODIS pixels (which, were, for these studies, square areas 500 m on a side) [105].

Interest in issues surrounding geometric misalignment between plots and pixels was particularly relevant for MODIS imagery, as its resolution was much closer to the scale of forest patches and

FIA plots than that of the unwieldy 1 km AVHRR pixels. This observation motivated research by Nelson et al. [107] into relationships between pixel size and correlations with FIA estimates; they found that the finest resolution tested (30 m pixels) led to estimates that aligned closest with those of FIA, whereas, surprisingly, larger pixel sizes (90–150 m) led to better estimates overall. That finding aligns with the recommendations of Hoppus et al. [89], who suggested that a 90–150 m pixel aggregation area is most appropriate for intersection of FIA plots with pixels; it also suggests that, despite the geometric incongruence, it was potentially feasible to use 250 m MODIS products with FIA data to meet business needs.

More research into the use of MODIS data for FIA business processes ensued. For example, Liknes et al. [108] assessed the value of MODIS for post-stratification of FIA plots to improve precision of forest area estimates, and found that MODIS imagery performed similarly to that of Landsat. However, Holden et al. [109] assessed the use of a 250 m pixel MODIS-based biomass map for stratification in the North Central FIA region and found little benefit compared to a simple random sampling approach with no stratification. Goeking and Patterson [110] describe the operational use of MODIS imagery for post-stratification of FIA data in the Rocky Mountain region. It is noteworthy that many of the above-mentioned studies show stronger benefits of RS integration for improving precision of forest area estimates, but smaller precision benefits for estimates of attributes like volume or biomass. FIA has traditionally used one stratification layer for both types of estimates, and further research is needed to determine the value of using different stratification layers for different attribute types.

2.6. Growth of Machine Learning

During this period of rapid increase in the availability of data from Landsat, MODIS and other sensors, a concurrent blossoming of computer storage and processing power was underway. These increased capabilities brought software tools from the field of what came to be called machine learning to the forefront in the RS community; model building and application with complex, nonlinear algorithms became accessible to the RS practitioner like never before. Previously, relatively simple classifiers, such as rudimentary supervised (based on multivariable nearest neighbor analysis) and unsupervised (based on automated cluster analysis) classification algorithms were commonly used, mostly because their implementation requirements were met by computing capabilities of the time [85]. The expansion of computer storage and processing power made more complex algorithms accessible to RS scientists, and as interest grew in improving environmental monitoring, scientists and statisticians from other fields were drawn to study RS and began to develop and apply techniques and algorithms such as those from the field of data mining and ecology [111].

FIA began to invest heavily in machine learning during the early 2000's. For example, Moisen and Frescino [112] integrated much of the institutional RS knowledge that FIA had accumulated up to that point in a study that combined FIA plots with AVHRR, TM, and digital topographic data to test the ability of several machine learning algorithms to produce maps of various forest characteristics for several states in the Western US. This research demonstrated the feasibility of applying previously-inaccessible algorithms to FIA data, and again highlighted that even after the turn of the 21st century, one of the main impediments to operationalizing these mapping techniques was computer processing speed.

One way these processing speed limitations were overcome was to use simpler algorithms and leverage existing software. For example, McRoberts et al. [113] implemented a k-nearest neighbor algorithm (k-nn), which is a machine learning technique based on multivariate similarity between pixel values associated with FIA plots and those without plots, to produce maps of FIA attributes. Implementing this technique is relatively straightforward in computer languages well-suited for matrix calculations, or through the adaptation of existing software packages such as Erdas or ArcGIS, as was done by Lister et al. [114] to produce k-nn-derived maps for New Hampshire. Ohmann and Gregory [115] used a k-nn-like approach with FIA data and Landsat to create maps of imputed estimates of forest attributes in Oregon. Nelson et al. [116] compared the stratification efficacy of forest/nonforest maps derived from NLCD with those from several machine-learning approaches, including k-nn.

New data processing tools and scripts streamlined workflows for combining plots and pixels and applying machine learning models to RS data. This included the development of freeware, such as spatial analysis functions for the R statistical software [117], as well as the adaptation of various commercial packages to work with RS data [118]; several of these were reviewed by Ruefenacht et al. [119]. These advances allowed for the achievement of a milestone for the FIA RS community: the production of the first national-scale map of forest biomass [120] and forest type [121]. The production of these maps demonstrated that it was feasible for FIA scientists and partners, working together, to achieve the production of national-scale maps integrating FIA data and RS observations.

A second major milestone for FIA was achieved by Wilson et al. [122], who produced national-scale maps of forest tree basal area by species using MODIS and other GIS data, and by Wilson et al. [123], in which nationwide maps of carbon stocks were produced. As with the work by Moisen and Frescino [112], these achievements were built using institutional knowledge acquired over the previous decades. The study built upon machine learning work done with FIA plots [113–115], accumulated experience with MODIS [107,120,121], and a series of past workflow development strategies for linking FIA plots, RS imagery, and various software platforms for GIS-based modelling. A similar effort was conducted by the USFS Forest Health Technology Enterprise Team (FHTET) [124]; the FHTET product was used in the production of the National Insect and Disease Risk Map [125], which is used by forest managers and policy makers to understand the potential impacts of forest pests. The role of FIA data in the production of the Blackard et al. [120], Ruefenacht et al. [121], Wilson et al. [122], and Ellenwood et al. [124] maps not only showed what could be done, but showed that these new techniques could be operationalized and made part of a suite of standard, FIA-based products.

2.7. Advanced Uses of Landsat

2.7.1. Opening of the Landsat Archive

The most important stimulus of research and development of Landsat-based science was the 2008 change in the U.S. government's Landsat Data Distribution Policy to allow for the release and download of the entire Landsat archive [126,127] at no cost to users. Coupled with enhancements in web-based RS data discovery and bulk ordering and download tools like USGS's Earth Explorer [128], obtaining Landsat data, which had previously been expensive and often irksome, had become nearly effortless. Similar mechanisms were developed by the European Union's Copernicus program [129], and nascent attempts to provide analysis-ready data, such as the Web-Enabled Landsat Data (WELD) [130], facilitated use and research of mosaicked, cloud-corrected, normalized imagery. Along these same lines, the Landsat archive was being pre-processed with various high-level algorithms like the Landsat Ecosystem Disturbance Adaptive Processing System (LEDAPS) to make data easier to use and facilitate retrospective analyses and monitoring studies [131,132].

2.7.2. Vegetation Change Tracker and the North American Forest Dynamics Project

The theoretical ability to obtain the entire Landsat record, pre-processed in useful ways, led to some ground-breaking research into what have become known as Landsat time series (LTS) for forest monitoring [133]. LTS consist of stacks of Landsat scenes, collected at a regular time interval over a long period, allowing for the tracking of the value of some image characteristic like a spectral index for each pixel and scene in the stack [134] (Figure 1). FIA became involved in nationwide use of LTS data for forest monitoring through participation in the North American Forest Dynamics (NAFD) project as a way to reduce spatial and temporal uncertainty in carbon estimates related to forest dynamics [135–138]. FIA data were used throughout the decade-long NAFD project to validate year of disturbance estimates from the LTS-based models [139], to refine spectral thresholds to detect trees in xeric systems [140], to interpret ratios of forest disturbance area to removal volumes across regions [137], to investigate annualized timber product outputs [141], to annually map the magnitude of harvests using repeat FIA basal area measurements [142], and to help attribute causal process to disturbance events [139,143].

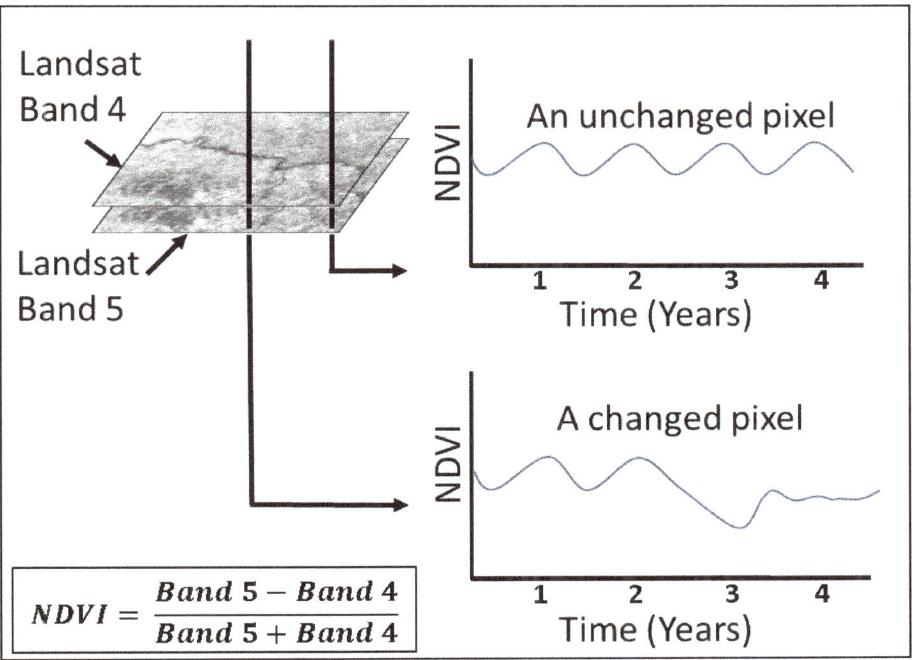

$$NDVI = \frac{Band\ 5 - Band\ 4}{Band\ 5 + Band\ 4}$$

Figure 1. Conceptual diagram of the creation of Landsat time series (LTS) analysis inputs. For each pixel, brightness values for a user-specified combination of layers is measured for each scene in the record (for Landsat 8, a new image is acquired every 16 days). Normalized Difference Vegetation Index (NDVI), a commonly-used vegetation index, can be calculated for each scene from Landsat 8's bands 4 and 5, and plotted against time. The properties of the resulting time series can be assessed to identify changes through time.

The Vegetation Change Tracker (VCT) algorithm [134,144], used to generate the NAFD LTS and annual change maps, was the first algorithm with nationwide application to reliably identify moderate through severe forest canopy loss events by detecting statistical anomalies in the LTS spectral signatures. These NAFD products, now archived at the Oakridge National Laboratory [136], allowed FIA projects to capitalize on annual LTS and change maps. For example, Stueve et al. [145], in order to improve annual estimates of forest change, adapted VCT by using information from snow-covered areas in the Lake Superior and Lake Michigan basins. For the same area, Garner et al. [146,147] then used the modified VCT data to improve the NLCD forest cover classes by subdividing them into successional stages, and Tavernia et al. [148] used these modified maps to characterize early successional habitat using FIA data. In a similar study, Nelson et al. [149] used VCT products to identify and characterize young forests in Wisconsin. Powell et al. [150] used NAFD and FIA data to quantify time series of live aboveground forest biomass dynamics. Brown et al. [151] used VCT products to assess impacts of surface coal mining on forest distribution in West Virginia. Schroeder et al. [152] investigated using FIA plot data, time series observations, and annual change maps to improve estimates of forest disturbance using model assisted post-stratified estimation. Moisen et al. [153] modified a statistical technique using constrained spline fitting and temporal shapes building off of NAFD data to model, map and monitor forest dynamics over three decades. A methodology for forest change process attribution, building off of NAFD LTS and an ensemble of change detection algorithms, was described and piloted in Schroeder et al. [143]. Dugan et al. [27] and Birdsey et al. [28] combined FIA data with VCT and similar outputs to characterize carbon dynamics by disturbance type for U.S. National Forests. In 2020, Schleeweis et al. [139] delivered nationwide maps attributing forest disturbance type.

2.7.3. TimeSync and LandTrendr

Two other tools were being developed by the LTS research community around the same time VCT was created: LandTrendr [154] and TimeSync [155]. LandTrendr (Landsat-based detection of Trends in Disturbance and Recovery) uses the then novel principles of exploiting LTS data to model trajectories of change, capitalizing on breaks in the trajectory trends (rather than statistical anomalies), which allow segments and vertices to capture and characterize long slow declines as well as abrupt losses. TimeSync is a visualization and database tool aimed at providing image interpreters with an environment and suite of tools to visually assess multi-band spectral signatures through time and record labels of land cover and land use for individual breaks (vertices) and trends (segments) over decades of imagery (radiometrically calibrating, cloud and shadow filling and chipping images along the way). FIA quickly saw the value of these tools, and the potential benefits of conducting LTS analyses on FIA plots to relate forest attribute data with temporal and spectral signatures from LTS. For example, Ohmann et al. [156] used LandTrendr to radiometrically correct Landsat scenes and to identify forest change as part of an old-growth forest change mapping effort. Schroeder et al. [152] demonstrated how manual interpretation of LTS imagery improved estimates of area and type of forest canopy disturbance in the Uinta Mountains of Utah; they collected a new variable (evidence of past disturbance) from the LTS using an approach modeled on TimeSync, and used maps derived partially from LandTrendr for post-stratification for variance reduction. Bright et al. [157] used LandTrendr to estimate characteristics of forests affected by bark beetle in five western states. Bell et al. [158] used an improved version of TimeSync [159] to study historic drought effects on forest canopy decline. Zhao et al. [140] used FIA data with the TimeSync tool to assess historic forest cover changes and their causes. Schroeder et al. [143] used TimeSync data, FIA plots, and outputs from VCT to classify forest disturbance causality in ten Landsat scenes distributed across the United States. Gray et al. [160] used TimeSync to identify and model historical changes in aboveground woody carbon on FIA plots. Filippelli et al. [161] used LandTrendr data with FIA plots to assess historical trends in pinyon-juniper biomass across parts of several western states. Research into the use of these technologies, as well as into the use of time series data from the Sentinel-2 mission for tracking forest change [162], is continuously evolving.

2.7.4. Landscape Change Monitoring System

The Landscape Change Monitoring System (LCMS) project, which grew out of the multi-agency Monitoring Trends in Burn Severity (MTBS) project [97], was a USFS-led initiative to exploit the LTS record by using machine learning algorithms to map landscape change. One of the main premises behind LCMS was that although different image classification algorithms will lead to slightly different maps, each with their own strengths and weaknesses, the ensemble of maps made from these algorithms can be summarized to improve upon any of the individual maps that constitute the ensemble. Schroeder et al. [143] and Cohen et al. [163] showed how using ensemble approaches for change detection improved upon algorithms that used individual algorithms and/or spectral indices; they combined several LandTrendr outputs in order to assess the impacts of various band and spectral index choices on classification accuracy. Schleeweis et al. [139] used an ensemble approach exploiting different algorithms for detection and classification of the type of disturbance processes occurring nationally, based on LTS analysis. Healey et al. [164] used an ensemble approach that leveraged the strengths and weaknesses of multiple algorithms, balancing their omission and commission errors, to create an improved overall model and map of forest change locations through time. Based on this initial work, an operational, cloud-based LCMS Data Explorer web application (http://lcms.forestry.oregonstate.edu/)

was created by the Forest Service. This tool was designed as a way to analyze, subset, and download LCMS data products, which include forest cover loss by year since 1984.

2.7.5. Use of LTS-Derived Covariates for Mapping of FIA Attributes

Additional exploration of the fusion of LTS and FIA data for FIA attribute mapping includes work by Wilson et al. [165], who used harmonic regression on LTS and FIA data for the state of Minnesota to map several FIA attributes; they found that the regression coefficients derived from Fourier analysis on LTS data improved mapping accuracy in important ways compared to alternatives. Brooks et al. [62] found that LTS-derived maps performed well as the source of strata for post-stratified estimation of FIA attributes. Derwin et al. [166] used harmonic regression coefficients from LTS in a canopy cover estimation approach, and found that they performed better than other LTS approaches when compared to FIA data. Research into creative uses of LTS for improving FIA business processes is in its early stages.

2.8. Cloud Computing

2.8.1. Cloud-Based Data Processing

One of the most significant advances in RS data analysis over the last several decades is the development of a cloud-based image processing and storage system called Google Earth Engine (GEE) [167]. GEE takes advantage of high-performance, parallel computing systems and the petabytes of RS data from several archives to create a cloud-computing environment for image storage and analysis. Previous workflows for large area mapping involved ordering, downloading (or mail ordering CDs or DVDs), pre-processing, mosaicking, and applying algorithms to often dozens of images on personal computers or local servers [85] (Figure 2). GEE removes several of these steps by hosting analysis-ready, mosaicked RS imagery data on distributed servers, and providing access to tools for pre-and post-processing, building models, and applying the models to the imagery. This has made large area mapping on a repeated basis accessible to a broad user community. Before FIA actively began using cloud computing, Hansen et al. [168] used GEE for forest mapping, creating yearly, global, LTS-based tree cover change maps between the years of 2000 and 2012 using machine learning algorithms applied to more than 650,000 Landsat scenes (20 terapixels of data processed using one million CPU-core hours). Since then, this approach has been operationalized and incorporated into an annual tree cover monitoring system called Global Forest Watch (GFW) [168,169]. GFW is particularly well-suited to detect change in tree cover in areas with continuous canopy cover.

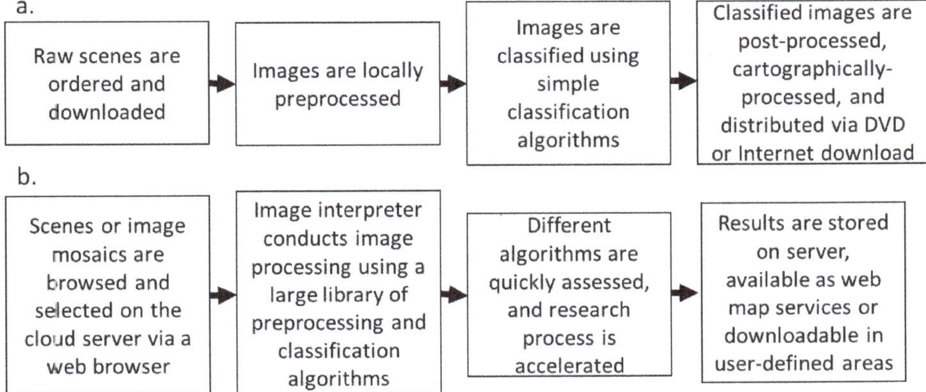

Figure 2. Description of two processes for conducting remote sensing. (**a**) Traditional approach, prior to era of cloud computing; (**b**) Workflow that is used during the era of cloud computing.

The trend of operationalizing change detection algorithms with GEE continued with subsequent conversions of TimeSync [155], LandTrendr [170], and LCMS ensemble model [164] code and workflows to GEE code, along with those of another algorithm, Continuous Change Detection and Classification (CCDC) [171]. CCDC uses a two-step cloud, shadow, and snow-masking algorithm, along with various metrics of spectral change from LTS data, to continuously detect change as new scenes are acquired. The use of CCDC algorithms on cloud computing platforms and supercomputers allowed for the full implementation of the U.S. Geological Survey Land Change Monitoring, Assessment and Projection (USGS LCMAP) initiative [172,173]. LCMAP, like the LCMS ensemble-based maps, seeks to provide an operational landscape change monitoring system; the approaches differ based on differences in the algorithms employed, the classification systems, and the fact that LCMS focuses mostly on forests.

2.8.2. Cloud-Based Data Hosting and Serving

The NAIP program's transition of CD- or DVD-based digital aerial imagery to online imagery was the harbinger of a paradigm shift toward serving massive datasets to users through web services. The most commonly-used web map services protocols are the Web Map Service (WMS), Web Feature Service (WFS), and Web Coverage Service (WCS) [174]. These types of map services allow client software, like a user's GIS, to interact with either pre-rendered images of the geographic entities, or the data associated with the geographic entities themselves. For example, if a user loads a local GIS data file of FIA plot locations, NAIP imagery from an Internet-based server such as the National Map (https://viewer.nationalmap.gov/services/) can be loaded beneath the plots so an interpreter can examine the imagery associated with each plot location. A WFS could also be used to load a vector-based watershed layer as well, so the user can assign attributes associated with each watershed to each plot. Finally, a WCS can be used to load Landsat imagery, and the pixel values of the imagery could be attached to the FIA plots and used for model building or validation. The map services are designed to optimize streaming of the GIS data to the client computer by returning only what is needed (the area and resolution) based on the context of the client GIS field of view; this minimizes latency and consumption of Internet bandwidth. Popular commercial Internet map services that rely on similar principles include Google (Google Maps and Google Earth) and Microsoft (Bing maps), and there are many open-source providers of map services [174] as well.

2.9. Increased Use of NAIP

2.9.1. Image-Based Change Estimation (ICE) and Logistical Planning Prior to Fieldwork (Pre-Field)

Map service technology allowed for several innovations within FIA and the broader Forest Service, particularly through the enhanced use of the NAIP imagery, which is often served through a WMS [72]. NAIP imagery generally consists of frequently-reacquired, nationwide, 3 or 4 band digital imagery with pixels with 1 m resolution [72]. Using NAIP, pre-field preparatory work became streamlined, with field crews being able to easily access recent imagery of plots online to determine if field visits are necessary and to obtain navigation and context information [175].

Efficient, rapid PI of high-resolution imagery in a GIS also became feasible. Toney et al. [176] compared results from PI vs. field-derived canopy cover estimates and found that the PI estimates were 10–20% larger than those in the field. Rapid PI was exploited by Lister et al. [177] for a trees outside forest (TOF) inventory for the states of North Dakota, South Dakota, Kansas and Nebraska, in which tens of thousands of photo plots were interpreted with respect to the presence or absence of TOF, and the plot labels were used to conduct prestratification to improve the efficiency of a ground survey. In a double sampling for post-stratification context, Westfall et al. [178] demonstrated the value of PI of WMS-based NAIP imagery for improving inventory efficiency in three counties of Pennsylvania. Frescino et al. [179] used digital photographs to conduct an inventory of TOF in Nevada. Lister et al. [180] used a rapid PI approach to monitor land use change in Maryland, and Lister et al. [181] used a similar method to characterize forest degradation in Maryland and Pennsylvania. Nowak and Greenfield [182–184] characterized land cover and urban tree dynamics at the national scale using rapid PI in Google Earth, and Nowak et al. [185] developed a tool called i-Tree canopy that allows users to create PI projects online using Google imagery as a backdrop. Based on similar principles, the FIA program developed a GIS tool using WMS-based imagery in order to collect canopy cover information for updates to the NLCD canopy cover project and other FIA business processes [186], and Lister et al. [180] developed a software tool for rapidly interpreting land cover on digital photo plots.

Building upon lessons learned from these and similar research efforts, the FIA program operationalized this technology for forest monitoring with the Image-based Change Estimation (ICE) project [187]. ICE was developed partly as a response to a 2009 resolution by the National Association of State Foresters that encourages FIA to increase the use of RS data for monitoring forest cover dynamics [188]. Furthermore, there was a perception among FIA partners that the FIA cycle length, which ranges between 5 and 10 years, was too great to identify important forest cover changes in a timely way. The ICE project addressed this by interpreting imagery associated with each new NAIP cycle, which is generally 3 years long [72]. Estimates include area by land cover, land use, and change category and agent, and are being used to address 2018 Farm Bill objectives of more timely production of estimates of forest trends and increased use of RS technology in monitoring.

2.9.2. Pixel-Based Mapping Using NAIP

Automated image classification of high resolution satellite imagery has been of interest in FIA for decades, but issues surrounding its large cost have limited its use. NAIP, on the other hand, is supplied by the USDA at no cost to users, and is therefore an appealing option [189]. Various attempts have been made to combine NAIP imagery with FIA plot information for pixel-based mapping. For example, Meneguzzo et al. [190] compared pixel-based and other classification methods using NAIP, and found that pixel-based classification performed similarly to an ocular photointerpretation method in Minnesota. Hogland et al. [191] used FIA plots as training data for classification of NAIP using a machine learning algorithm, and found that using the NAIP-based map improved estimates of various FIA attributes compared to using plots alone. Hogland et al. [192] used NAIP and its derivatives with FIA data to tree density and basal area for portions of Georgia, Alabama and Florida, and cited the need to download NAIP imagery for local processing as being one time-consuming element of the project. To circumvent this challenge, Chang et al. [193] streamlined pre-processing and acquisition of NAIP imagery by using the cloud-based GEE platform in a project that relied on machine learning to map several FIA attributes in California and Nevada.

2.9.3. Object-Based Image Analysis Using NAIP

In addition to manual PI, FIA has used NAIP imagery for semi-automated land cover class mapping using object-based image analysis (OBIA) [194,195] (Figure 3). OBIA works by using algorithms that assign each pixel in an image to a spatially-contiguous cluster, typically a polygonal area that meets a homogeneity criterion used as a parameter in the algorithm. This in effect creates a map of land cover polygons that can then be classified using machine learning approaches. Frescino et al. [189] compared OBIA-based maps produced using NAIP and FIA data with those produced with other high resolution imagery and found that it performed nearly as well as the more costly imagery. Lister et al. [196] fused Landsat and NAIP imagery and performed OBIA to impute forest inventory information to stands in a forest in Maryland. Liknes et al. [197] used NAIP OBIA-based machine learning algorithms to classify tree cover in agricultural areas in North Dakota. Riemann et al. [198,199] assessed the value of using OBIA-based canopy detection products, which partly-relied on NAIP, for estimating canopy cover in several states. Meneguzzo et al. [190] and Meneguzzo [200] found that image segmentation compared favorably with or improved upon pixel-based classification of NAIP imagery to map trees outside of forest. Liknes et al. [201] used a NAIP OBIA-derived map of tree cover patches and a patch shape detection algorithm to identify windbreaks in Nebraska, and Paull et al. [202,203] and Kellerman et al. [204], produced statewide, NAIP-based maps of tree canopy for both Kansas and Nebraska. These studies show the potential for using the spectral information contained in the multi-band NAIP imagery, but do not fully exploit the structural information that could be obtained from stereo photogrammetric analysis of overlapping NAIP images.

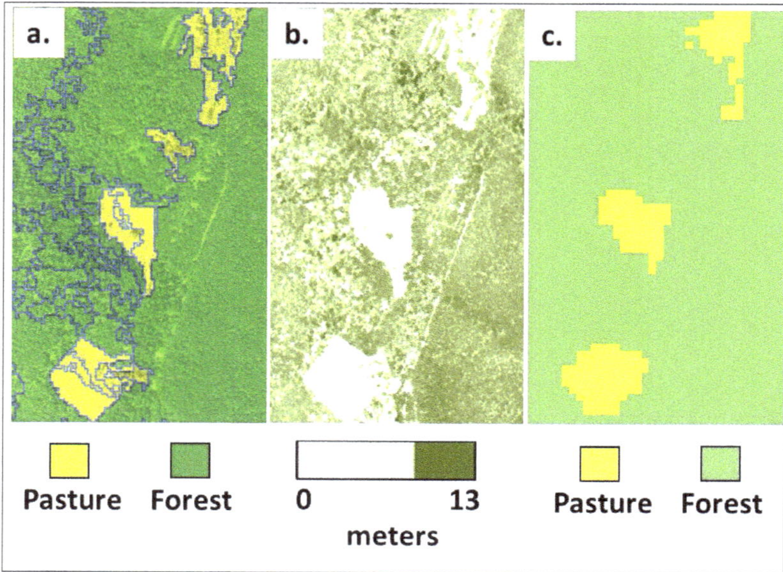

Figure 3. Comparison of different uses of National Agriculture Imagery Program (NAIP) for forest resource measurement and mapping with Landsat-based classification. (**a**) A NAIP image with an image segmentation applied (blue polygonal areas) and an object-based image analysis (OBIA) classification algorithm applied to the polygons (yellow is nonforest, green is forest); (**b**) a pixel-based tree height classification of NAIP imagery; (**c**) A Landsat-based forest/nonforest classification.

2.9.4. 3-D Processing of NAIP for Structure

New technology in software and processing allows for the development of optical 3-d models from stereo NAIP imagery [205]. The approach works by using advanced software to process digital aerial photography of the same location taken from different angles to create 3-d point clouds that represent the height of objects (such as tree canopies). The advantage of this approach over active RS technology like Light Detection and Ranging (lidar, described in the next section) is that it can be over an order of magnitude smaller in cost [206]. A disadvantage over lidar is that the quality of the information from a NAIP-derived point cloud is typically lower. FIA therefore performed several tests to determine whether derivatives of 3-d NAIP imagery are of value to enhance forest inventory. For example, Gatziolis [207] obtained promising results when using texture metrics from digital aerial photography for estimating canopy structure parameters in a range of forest types in Michigan. Webb et al. [205] investigated the value of NAIP point clouds compared to those from lidar for vegetation mapping and found that they were comparable in certain conditions, but that NAIP in general was not as accurate as lidar. Strunk et al. [206] created NAIP-based canopy height models under different processing scenarios and determined that using the height maps in FIA estimation process for stratification led to a 400% increase in precision of volume estimates over models without RS-based strata.

2.10. Airborne Light Detection and Ranging (Lidar)

Lidar methods have revolutionized several aspects of forest inventory over the past several decades. In a lidar system, a laser, typically mounted on an airplane, emits a beam of light, the transmission time between emission of the light, reflection of the light from an object of interest, and its return to the receiving sensor is recorded, and the time data are converted to information about the structural characteristics of the object that reflected the light [208]. There are various types of lidar data types, and they have been used with FIA data for wall-to-wall mapping, sample-based estimation, and field measurement of forest characteristics.

2.10.1. Airborne Lidar for Wall-to-Wall Mapping

FIA data have been used in concert with airborne lidar in various ways for many years. Relationships between the two data sources have been explored by Gatziolis [209,210] and Schrader-Patton [211], who demonstrated that lidar can be used to refine the locations of FIA plots by matching individual tree locations. Gatziolis [212] used lidar to create maps of Site Index (an index related to tree growth potential) and later [210] compared canopy cover estimates from an FIA PI approach with those from lidar and found systematic overestimation for small canopy cover values and underestimation for large canopy cover values. Riemann et al. [198,199] assessed the utility of lidar for replacing PI as a way to estimate canopy cover, and found that there were important differences between PI, field-mapped, and OBIA-based, lidar-dependent estimates, particularly in areas with small amounts of canopy cover. Andersen et al. [213] evaluated the error budgets in lidar-derived estimates of tree height for deciduous species and conifers, while Li [214] and Gopalakrishnan et al. [215] found that the correlation between lidar-derived tree heights and those measured by FIA was high. However, Gatziolis et al. [216] documented discrepancies between lidar-derived and field inventory estimates of tree height in challenging U.S. Pacific Northwest conditions. Possible reasons for the discrepancies found in these studies include field-lidar georeferencing mismatch, variable lidar quality, and field crew method inconsistency.

Due to the correlation of lidar-derived information with forestry attributes of interest, lidar data have often been used as a covariate in predictive modeling studies. The structural information contained in the lidar data can help inform models of forest volume or biomass, canopy cover, and land cover class. One approach to this is to use wall-to-wall maps of lidar as the main input to models. For example, Skowronski et al. [217] used FIA data and lidar in New Jersey to map forest structure and fire fuel loads. Johnson et al. [218,219] used lidar in concert with FIA plots to make high-resolution maps of forest carbon in Maryland. Sheridan et al. [220] summarized lidar data on and in the vicinity of FIA plots and modeled and mapped tree volume and biomass in Oregon. For a study site in Hawaii, Hughes et al. [221] conducted the first study to map aboveground carbon across a tropical landscape with lidar and FIA information. Joyce et al. [222] installed and measured FIA-like plots for producing lidar-based models of coarse woody debris in Minnesota.

Studies over large areas, such as U.S. states, where lidar data are the sole predictor used in models of FIA attributes are relatively rare because lidar data are massive, and significant pre-processing is required to generate an analysis-ready dataset for spatial modelling over these large areas. Airborne lidar data have traditionally been more expensive than optical imagery, and repeat acquisitions supporting assessment of forest dynamics or other forest characteristics that are linked to phenology are infrequent. It is therefore much more common for lidar to be used in concert with other data sources such as Landsat when mapping over large areas. For example, Lefsky et al. [223] combined FIA plot data, lidar, and Landsat to map forest biomass in Oregon and Washington. In a study combining various RS data types and FIA data, Chopping et al. [224] fused MODIS and data from NASA's Multiangle Imaging Spectroradiometer (MISR) to produce maps of biomass and forest structure in the southwestern U.S., and used FIA-based maps and lidar data to assess results. Deo et al. [225] compared LTS, lidar and fused LTS-lidar datasets for back-projecting biomass to a baseline year of 1990, and found that fusion of LTS and lidar improved results over alternative models. In similar studies that fused lidar, Landsat, and FIA data in four states in the eastern U.S., Deo et al. [226] assessed the advantages of using site-specific models vs. generic models of forest biomass, Deo et al. [227] assessed the impacts of spatial resolution of RS data on predictive accuracy, and Ma et al. [228] compared the performance of models using various combinations of the LTS, lidar, and FIA inputs. Bell et al. [229] used FIA data to compare lidar and LTS maps of aboveground biomass in 3 regions in Oregon and Washington, and found that lidar performed better than LTS. Gopalakrishnan et al. [230] also used LTS and lidar data, but fused them to impute site index estimates to loblolly pine stands over large areas in the southeastern U.S.

2.10.2. Airborne Lidar for Sample-Based Estimation

FIA's main source of estimates for reporting is currently not maps of forest attributes (what is colloquially referred to as "pixel counting"), but rather the information from its large network of ground plots. This is due partly to its legislative mandate, historical reliance on traditional survey sampling, and need for consistently reliable long-term data, and partly to the lack of practical ways to quantify systematic error (bias) in ways that are agreed upon by the scientific community [5,33]. However, lidar can also be collected in a sampling mode and be used practically for model-assisted inference, as described above. For example, Andersen et al. [231] used structural measurements from airborne lidar strip sample-based maps made from model-assisted regression as the second stage in a two stage estimation procedure in Alaska. Alonzo et al. [232] demonstrated an application of repeated measurements of airborne lidar samples to assess fire effects over FIA plots in Alaska. Strunk et al. [233,234] demonstrated a regression estimation approach using inventory data and lidar from a study site in Washington. McRoberts et al. [42] employed lidar-assisted regression estimators with FIA data as part of a comparison of different approaches at a study site in Minnesota. In a similar study over the same area, McRoberts et al. [235] assessed the "shelf life" (applicability through time) of lidar data that were not collected contemporaneously with inventory data when conducting model-assisted estimation. All of the applications mentioned in this section so far are based on airborne lidar; however, spaceborne lidar systems can be used in forest inventory applications as well.

2.11. Spaceborne Lidar for Sample-Based Estimation

2.11.1. GLAS

One of the objectives of the Ice, Cloud, and land Elevation Satellite (ICESat), which carries the Geoscience Laser Altimeter System (GLAS), is to measure vegetation canopy height [236,237]. GLAS data consist of sets of crossing transects that cover much of the Earth, with detailed laser height measurements taken on points distributed along these transects. Lefsky et al. [238] found good agreement between canopy height measurements from GLAS and field-measured heights from modified FIA plots in Tennessee and Oregon, and Pang et al. [239] found strong relationships between canopy height estimates from GLAS, airborne lidar and field measurements on modified FIA plots. Plugmacher et al. [240] found a somewhat weak relationship between FIA and GLAS-based height data in the Appalachian mountains, as did Li et al. [241], who used a GLAS-Landsat fusion approach in young forests in Mississippi. None-the-less, GLAS data have been shown to be useful for measuring not only height, but also forest biomass relative to estimates from FIA data in boreal forest regions of Alaska [242], and, in a unique fusion of FIA and other inventory data, airborne lidar, and GLAS, for the conterminous U.S. and Mexico [243].

2.11.2. GEDI

A problem identified by Nelson et al. [243] with the GLAS approach for estimating biomass or other attributes regionally is that calculating confidence intervals is not straightforward due to the combination of models used in the estimation procedure. In anticipation of this challenge, Healey et al. [244] proposed a sampling design that allows for the use of FIA data and nearby GLAS data for biomass estimation, using the state of California as a pilot study. Other approaches were also developed in anticipation of the GLAS-like data provided by NASA Global Ecosystem Dynamics Investigation (GEDI) mission, which includes a full-waveform lidar instrument that was mounted on the International Space Station (ISS) and was launched in 2018 [245]. For example, Ståhl et al. [59], McRoberts et al. [58], Saarela et al. [246], and Patterson et al. [247] discuss estimators that combine FIA and GEDI data. They use various combinations of wall-to-wall optical data (typically Landsat), a sample of more highly-correlated data (such as GLAS or GEDI), and sparse ground plots (FIA) to produce estimates and confidence intervals that are interpretable through the lens of sampling theory. Similarly, Nelson et al. [243] use a multi-phase sample that combines ground, airborne lidar and spaceborne lidar to generate estimates of biomass that are compatible with those from survey sampling. GEDI and similar datasets like those from the next generation of the ICESat mission (launched in 2018) offer opportunities to enhance or replace certain monitoring activities currently conducted with ground inventory plots, due to the lower overhead of sample-based methods and frequent reacquisition of data.

2.12. Unmanned Aerial Systems and Terrestrial Lidar

FIA is currently in the early investigatory phase of research into operational use of Unmanned Aerial Systems (UAS), which consist of aircraft such as drones that carry some combination of active and passive sensor systems for imaging forests at the local scale. For example, Gatziolis et al. [248] used a UAS to develop 3-d models of individual trees and found that photo-based models compared well to more detailed lidar-based 3-d models developed separately. Fankhauser et al. [249] concluded that their use of UAS to measure tree heights and counts showed promise for supporting traditional forest inventories. The largest potential for future use of UAS operationally in FIA is likely their use in remote or inaccessible areas, such as was done by Alonzo et al. [250], who flew drones over FIA plots in Alaska and found that they were moderately effective at estimating forest type, basal area, tree density, and biomass.

Terrestrial lidar is another new technology in its early stages of exploration by FIA. Gatziolis et al. [248] and Strigul et al. [251] compared terrestrial lidar-based tree structure measurements with those from other sensors for testing 3-d tree imaging technologies, and Gatziolis et al. [252] and Klockow et al. [253] used it to gather data related to live and dead tree allometry, respectively. Both found that it offered important improvements over standard FIA methods, and showed potential for improving fieldwork efficiency. Once algorithms are refined and processing software and hardware are adequate, terrestrial lidar and UAS technology will have the potential to improve FIA's plot-based data collection efforts by collecting new structure attributes and eliminating certain manual measurements.

3. General Observations on RS Data Integration in FIA and Other NFIs

General Characteristics of FIA's Use of RS

There are several instructive principles that emerge when examining the evolution of FIA's use of RS technology:

- **NFI data are invaluable to creating RS products.** They provide a standardized source of training data for models, and their use raises the likelihood that RS-based estimates will align with NFI-based estimates. They also provide valuable validation data for users interested in conducting map accuracy assessments at both the plot-pixel scale, as well as over larger geographic areas like U S. counties, for which NFI-based estimates and confidence intervals can be generated.
- **A successful RS program has access to RS data inputs, software, and hardware, including affordable high performance computing systems.** There was a strong correlation between advances in FIA's use of RS and improvements in Internet and personal computer technology, and, more recently, a similar increase in RS technology usage with the opening of the Landsat archive, the advent of other free RS data input sources, and the advent of cloud computing systems. It cannot be understated how the democratization of RS data acquisition and processing

technologies have led to improvements in our ability to monitor forest resources, and how FIA scientists are contributing more and more to both basic and applied research aimed at advancing forest science in these areas.

○ **Advances in RS usage require nimbleness and outlets for creative investigation**. Support for intellectual fora such as program meetings and scientific conference attendance advances what McRoberts [254], citing Reichenbach [255], calls the "discovery" component of science, i.e., the exploratory and creative part of the scientific method that focuses on identifying research questions, forming hypotheses, and developing models. Mechanisms for scientists and technical staff to conduct research and share preliminary results in a less-formal way furthers advancements.

○ **Advances in RS are incremental, beginning with discovery and leading to operationalization**. Figure 4 is a conceptual model showing the process that FIA RS research has typically gone through over the last several decades, beginning with knowledge discovery and ending in operationalization. It is noteworthy that some of the studies described in this review have not yet, or never will, become operational; Figure 4 identifies several points in the research and development process where operationalization can be impeded:

(a) After research into methods for application is conducted, it becomes clear that it is not feasible, or results are not as expected due to poorly-conceived research ideas that attempt to integrate components of many studies and stakeholder needs.

(b) After prototype development, large costs of operationalization or a lack of research maturity may limit adoption likelihood.

(c) After operationalization of the technology, it becomes clear that the user community does not yet have the capacity to use the results of the new technology. Strategies to address this include continuous capacity building among the user community, continuous improvement of the technology, and technology transfer.

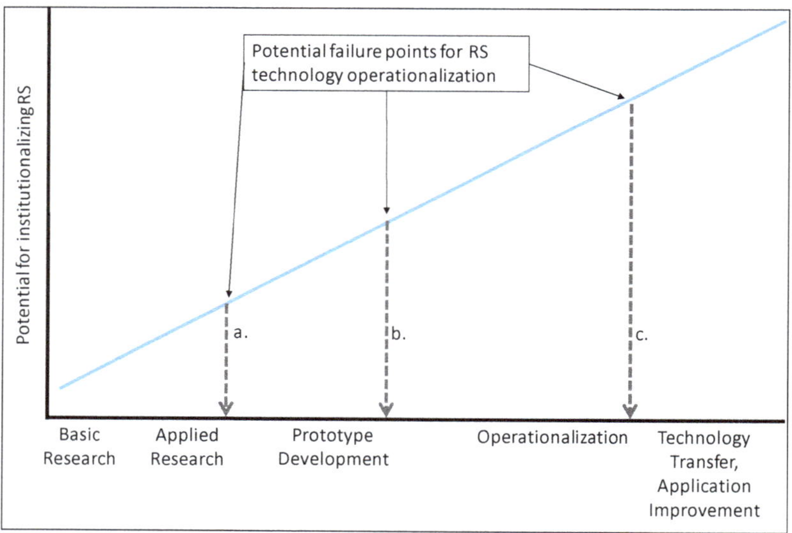

Figure 4. Conceptual graphic depicting the steps in a process in which the use of a new technology becomes institutionalized in Forest Inventory and Analysis (FIA). There are several points at which adoption of the technology might fail, including after applied research has been attempted (**a**), after prototype development (**b**), or after operationalization (**c**).

This case study reveals that FIA has created an environment where the competition of ideas and a culture of collegiality has led to creative thought and improvements in program efficiency, as well as a strong foundation of institutional knowledge that serves as a platform from which future research will advance.

4. Future Directions of RS Technology in FIA

In this section, several likely uses of new RS technology that could lead to increased operational efficiency in FIA or contribute to other advancements in forest monitoring are described.

4.1. RS Imagery Time Series

Use of FIA data in NAFD, LCMS, LCMAP, and more foundational work in LTS has set a clear course toward its future operationalization in FIA. Enhanced and novel LTS-based change detection algorithms are being explored in the FIA RS community (e.g., [256]), and opportunities for using data from other sensors, such as those from the European Union's Sentinel mission, with FIA data are in the knowledge discovery phase [162]. Analysis-ready LTS data are being created with new processing tools, making them more usable for operational monitoring [257]. It is likely that adopting LTS products will result in an increase of time series data use in mapping not only of change, but also of other FIA attributes (e.g., Wilson et al. [165]), and this will only become easier as more cloud-based options for computing and storage emerge.

Finally, in addition to LTS mapping applications, sample-based LTS approaches will be important in the future. For example, the current version of TimeSync [155] generates LTS graphics to aid staff performing sample-based PI, and Lister and Leites [162] showed how time series data associated with sample points can be used with machine learning to accurately identify forest cover change. Adopting a sample-based paradigm paves the way for its use to pre-screen inventory plots, which especially makes FIA's ICE project more efficient by identifying only the subset of FIA plots that need interpreting between ground visits. Moisen et al. [258] found that in order to report on land use and land cover trends in north central Georgia with adequate precision and temporal coherence, data needed to be collected on all the FIA plots each year over a long time series and broadly collapsed LULC classes. Relying heavily on TimeSync concepts, they call for a new, harmonized data collection approach that integrates ground, Landsat, and aerial imagery via a single enhanced plot interpretation process. Other FIA applications of sample-based LTS data include its potential for use to impute missing data at FIA plot locations that are unable to be measured in the field [259].

There are inherent advantages to using time series data as a sampling tool as opposed to a wall-to-wall mapping tool; maps require large amounts of additional resources in terms of storage, processing, and time spent on cartography and correcting visual discontinuities associated with image seamlines, clouds, and shadows. However, LTS data associated with sample points are more flexible, manageable, and can be linked to FIA plots for near real-time updates without the overhead of producing maps. Many natural resource monitoring programs around the world are turning to manual sample-based methods for estimating forest cover change, and many new efficient ocular PI tools have been developed [260].

4.2. Cloud Computing and Storage

Cloud and super-computing will clearly play a stronger role in the future of RS image processing, particularly when working with LTS or large volumes of high resolution imagery. As experienced with LandTrendR [170] and LCMS [261], transitioning to cloud computing in GEE opened possibilities for large area mapping that were previously not feasible. Similarly, supercomputers helped produce NAFD products [139]. GEE-based LTS advances include methods coming out of the Global Land Cover mapping and Estimation (GLanCE) project, which seeks to use GEE with the CCDC algorithm [171] to map land cover and land cover change annually at the global scale [262], as well as the recent addition of NAIP to GEE, which will open up the potential for large area, high resolution pixel- or OBIA-based mapping using NAIP imagery every few years.

There are several new open source and commercial cloud-based RS environments and software tools being developed in addition to GEE [263]. For example, FIA's BIGMAP project [264] is based on a cloud architecture that includes Amazon Web Services, Esri's ArcGIS Enterprise Image Services, and ArcGIS applications, and seeks to leverage imagery and FIA plot data for improved estimation, mapping, data analysis and data distribution to FIA staff and clientele. The advantage of the BIGMAP project is that it is integrated with the Esri software ecosystem, in principle making it easier for GIS specialists to use. It also allows for the use of FIA's plot coordinate data in a corporate environment while keeping private information protected, exploitation of cloud-computing, and integration with various statistical applications and data types.

4.3. Exploitation of the Z-Dimension

The "Z-dimension" is a colloquial term that refers to data that confer information on structure above the ground, as opposed to that from the X-Y plane, which is typically associated with optical RS data. Airborne and spaceborne lidar, and advances in the use of radar, will provide new opportunities to increase FIA's efficiency and improve mapping quality.

4.3.1. Airborne Lidar

Airborne lidar will likely occupy a growing role in FIA. Its unique value to FIA for work in remote areas of Alaska has been demonstrated [265], and as with LTS, it will become feasible for large area mapping as cloud-based technology improves and data discovery and distribution are standardized [215,266]. Furthermore, the USFS is currently conducting tests on lidar sensor systems that are combined with camera systems that could be used on future NAIP missions [267]; this would bring costs down, increase standardization, and shorten lidar revisit cycles, which at current use do not meet FIA's monitoring requirements. Large area lidar datasets could serve as covariates in predictive models of FIA attributes and forest cover change, and could serve as a means by which to help better-identify trees outside of forest and expand FIA tree data collection across all land types.

4.3.2. Spaceborne Lidar

Structural information from ICESat and GEDI lidar systems could take on added importance as data become available and FIA's understanding of the data grows. However, a barrier to its adoption is the lack of understanding of how to integrate sample-based lidar that only has limited spatial overlap with FIA plots. New estimation approaches that combine field plots, lidar point sample locations, and wall-to-wall lidar or other RS data will be needed, along with appropriate estimators that produce results compatible with FIA estimates [246,247]. Efforts are currently underway to create a web-based estimation system that does just that: The Online Biomass Inference Using Waveforms and Inventory (OBI-WAN) project, which seeks to provide a carbon reporting system that uses GEDI, FIA data, and novel estimators [246] to produce reports for user-specified areas. Similar approaches could be adopted with the ICESat2 data that became available in 2020 [268]; however, it should be noted that both GEDI and ICESat2 have finite mission lives, and there is no guarantee of data stream continuity into the future.

4.3.3. Radar

FIA has not conducted much research using radar data. A notable early application of radar exploits the InSAR (Interferometric Synthetic Aperture Radar) system carried on the Shuttle Radar Topography Mission (SRTM). With data from FIA plots, inSAR, and other imagery and GIS layers, Walker et al. [269] and Kellndorfer et al. [270] produced vegetation height maps for large sections of Utah and several northeastern states, respectively. Using FIA plot data, InSAR, and data from other sensors, Yu et al. [271] produced not only forest height but also biomass maps for the state of

Maine. Kellndorfer et al. [272] describe various realized and potential applications of using radar for large area mapping. Future work with SAR data for large area mapping will likely be conducted using cloud-based processing. For example, Google Earth Engine now hosts a long time series of Sentinel-1 SAR data, as well as algorithms for conducting change analyses with these [273]. As radar processing becomes more streamlined, such as in ways identified by resources such as Flores et al. [274], FIA scientists will be more likely to more fully adopt this promising technology.

4.4. Improved Estimation

A primitive, though unrealized vision for operationalization of a nascent form of SAE was presented by McRoberts and Wendt [275]. They advocated an approach where FIA clientele could identify arbitrary polygonal areas within-which FIA estimates could be generated using post-stratification information aggregated to cells in a tessellation of the country. In an unrelated effort, Proctor et al. [276] describe a well-known carbon tool, the Carbon OnLine Estimator (COLE), which did allow users to identify custom areas for which FIA-based carbon estimates were produced (also see Van Deusen and Heath [277]. After running for over 15 years, the COLE tool is currently not available as it is being updated for a cloud computing environment. More recently, new RS technology and increased computing resources allow FIA the opportunity to greatly improve estimation efficiency moving forward. A significant step toward operationalizing the use of alternative model-assisted inference is the development of the Forest Inventory Estimation for Analysis (FIESTA) software (Frescino et al., 2012), which currently has a model-assisted module that accesses the seven model-assisted estimators used in the statistical package mase [278]. FIESTA allows for not only the use of traditional FIA estimators that are compatible with corporate systems, but also the integration of estimators and data from machine learning for comparison with standard methods.

Moving beyond model-assisted inference, there is an increased need for small area estimates to characterize forest attributes within small domains such as counties, ecological boundaries, disturbance events, and shortened time intervals (i.e., SAE). Efforts to operationalize SAE have generally been hindered by a combination of technical challenges and reconciliation with FIA's standardized, peer-reviewed estimation system [14]. However, recently software developments and improvements in knowledge of machine learning and model-based inference have led to the development of new tools and demonstration projects. For example, the FIESTA software mentioned above also takes a step toward operationalizing the use of SAE methods by offering a module that easily links FIA and remotely sensed data through packages in the R software [279], including sae [280] and JoSAE [281]. Future developments in this area include a revision of FIA's estimation procedure handbook [14] to include SAE, improved model-based inference (e.g., Saarela et al. [246]), and tool creation to streamline adoption of novel techniques as they are developed and their value as part of FIA's standard operating procedures is established.

One possible step toward improving estimation in a way similar to the functionality of the COLE tool is the creation, through advanced k-nn techniques, of a national, pixel-based map of FIA plot identifier in the aforementioned BIGMAP platform that is under development. This national map, which would be driven in part by LTS data that are constantly refreshed, would serve as the foundation for summarization of FIA plot data at the pixel-level, allowing for the drawing of arbitrary polygonal areas and the calculation of estimates using either model-assisted or model-based inference [42]. A similar approach to this was already implemented by Wilson et al. [122] using MODIS data, and a Landsat-based FIA plot identifier map has been produced for the entire country for one time period using the technique described in Riley et al. [282]. The BIGMAP platform has the advantage that it can easily integrate with other FIA estimation infrastructure and databases, and thus potentially serve as a foundational platform for FIA's future operational use of SAE. Another possible step forward is to integrate FIESTA's statistical estimation capabilities with BIGMAP's powerful spatial toolsets to access a multitude of SAE tools already available on the Comprehensive R Archival Network (CRAN). Investigations are currently underway to identify best models and delivery systems.

5. Conclusions

FIA has a long history of using RS imagery and technology to improve efficiency. Beginning with hand-drawn maps, paper aerial photos, and coarse-resolution satellite imagery, FIA's use of RS has tracked advancements in technology, as it has tracked the progression of inventory data needs from solely-timber focused to more broadly ecological and social. It has evolved to exploit time series of satellite data such as Landsat, MODIS and Sentinel, and has been turning toward technologies that are linked to Z-dimensional structural information (height), such as lidar, stereo NAIP-derived point clouds, and radar. Technology advancements, including software, hardware, and networking capabilities, have made machine learning and cloud computing foundational for FIA's future RS advances, and led to a blossoming of efficiency-creating methods.

Barriers to operationalizing new technologies still exist, and include balancing research and core production functions, limited capacity and high costs to use new tools and data types, the need to maintain long-term continuity of results even as methods and outputs evolve, and use of production workflows that are difficult to change due to their complexity. However, expectations from stakeholders continue to encourage increased RS technology development. In addition, FIA has created an environment where the competition of ideas and a culture of collegiality has led to creative thought and improvements in program efficiency, as well as a strong foundation of institutional knowledge that serves as a platform from which future research will advance. NFIs such as FIA require a commitment to science, public service, and stewardship of institutional knowledge to evolve along with advancements in RS and natural resource monitoring science, while ensuring long-term comparable results.

Author Contributions: Conceptualization, A.J.L. and L.S.H.; writing—original draft preparation, A.J.L.; writing—review and editing, A.J.L., H.A., T.F., D.G., S.H., L.S.H., G.C.L., R.M., G.G.M., M.N., R.R., K.S., T.A.S., J.W., and B.T.W.; visualization, A.J.L.; project administration, A.J.L. All authors have read and agreed to the published version of the manuscript.

Funding: This research received no external funding.

Acknowledgments: This work was supported by the U.S. Department of Agriculture, Forest Service. The authors thank the contributions of a multitude of FIA staff and partners who have contributed to the research described herein.

Conflicts of Interest: The authors declare no conflict of interest.

References

1. Food and Agriculture Organization (FAO). *Global Forest Resources Assessment 2020—Key Findings*; FAO: Rome, Italy, 2020; Available online: http://www.fao.org/3/CA8753EN/CA8753EN.pdf (accessed on 13 December 2020).
2. Kangas, A.; Maltamo, M. *Forest Inventory: Methodology and Applications*; Springer: Heidelberg, The Netherlands, 2006.
3. Tomppo, E.; Gschwantner, T.; Lawrence, M.; McRoberts, R.E. *National Forest Inventories: Pathways for Common Reporting*; Springer: New York, NY, USA, 2009.
4. Shiver, B.D.; Borders, B.E. *Sampling Techniques for Forest Resource Inventory*; John Wiley and Sons: New York, NY, USA, 1995.
5. McRoberts, R.; Tomppo, E. Remote sensing support for national forest inventories. *Remote Sens. Environ.* **2007**, *110*, 412–419. [CrossRef]
6. McRoberts, R.E.; Tomppo, E.O.; Næsset, E. Advances and emerging issues in national forest inventories. *Scand. J. For. Res.* **2010**, *25*, 368–381. [CrossRef]
7. McRoberts, R.E.; Næsset, E.; Sannier, C.; Stehman, S.V.; Tomppo, E.O. Remote sensing support for the gain-loss approach for greenhouse gas inventories. *Remote Sens.* **2020**, *12*, 1891. [CrossRef]
8. USDA Forest Service. *Forest Inventory and Analysis Strategic Plan, FS-1079*; United States Department of Aagriculture Forest Service: Washington, DC, USA, 2016.
9. Nelson, M. Forest inventory and analysis in the United States: Remote sensing and geospatial activities. *Photogramm. Eng. Remote Sens.* **2007**, *73*, 729–732.

10. Barrett, F.; McRoberts, R.E.; Tomppo, E.; Cienciala, E.; Waser, L.T. A questionnaire-based review of the operational use of remotely sensed data by national forest inventories. *Remote Sens. Environ.* **2016**, *174*, 279–289. [CrossRef]

11. McConville, K.S.; Moisen, G.G.; Frescino, T.S. A tutorial on model-assisted estimation with application to forest inventory. *Forests* **2020**, *11*, 244. [CrossRef]

12. Gillespie, A. Rationale for a national annual forest inventory program. *J. For.* **1999**, *97*, 16–20.

13. Oswalt, S.N.; Smith, W.B.; Miles, P.D.; Pugh, S.A. *Forest Resources of the United States, 2017: A Technical Document Supporting the Forest Service 2020 RPA Assessment, WO-GTR-97*; U.S. Department of Agriculture, Forest Service, Washington Office: Washington, DC, USA, 2019.

14. Bechtold, W.A.; Patterson, P.L. *The Enhanced Forest Inventory and Analysis Program—National Sampling Design and Estimation Procedures, GTR-SRS-80*; U.S. Department of Agriculture, Forest Service, Southern Research Station: Asheville, NC, USA, 2005.

15. Burrill, E.A.; Wilson, A.M.; Turner, J.A.; Pugh, S.A.; Menlove, J.; Christensen, G.; Conkling, B.L.; David, W. *The Forest Inventory and Analysis Database: Database Description and User Guide for Phase 2 (Version 8.0)*; U.S. Department of Agriculture, Forest Service: Washington, DC, USA, 2020.

16. Piva, R.J.; Walters, B.F.; Haugan, D.D.; Josten, G.J.; Butler, B.J.; Crocker, S.J.; Domke, G.M.; Hatfield, M.A.; Kurtz, C.M.; Lister, A.J.; et al. *South Dakota's Forests 2010*; U.S. Department of Agriculture, Forest Service, Northern Research Station: Newtown Square, PA, USA, 2013.

17. Butler, B.J.; Crocker, S.J.; Domke, G.M.; Kurtz, C.M.; Lister, T.W.; Miles, P.D.; Morin, R.S.; Piva, R.J.; Riemann, R.; Woodall, C.W. *The Forests of Southern New England, 2012, NRS-RB-97*; U.S. Department of Agriculture, Forest Service, Northern Research Station: Newtown Square, PA, USA, 2015.

18. USDA Forest Service. *Future of America's Forest and Rangelands: Forest Service 2010 Resources Planning Act Assessment, WO-GTR-87*; U.S. Department of Agriculture, Forest Service, Washington Office: Washington, DC, USA, 2012.

19. USDA Forest Service. *Future of America's Forests and Rangelands: Update to the 2010 Resources Planning Act Assessment, WO-GTR-94*; U.S. Department of Agriculture, Forest Service, Washington Office: Washington, DC, USA, 2016.

20. USDA Forest Service. *National Report on Sustainable Forests—2003, FS-766*; U.S. Department of Agriculture, Forest Service, Washington Office: Washington, DC, USA, 2004.

21. Robertson, G.; Gualke, P.; McWilliams, R.; LaPlante, S.; Guldin, R. *National Report on Sustainable Forests-2010, FS-979*; U.S. Department of Agriculture, Forest Service: Washington, DC, USA, 2011.

22. Rudis, V.A. *Comprehensive Regional Resource Assessments and Multipurpose Uses of Forest Inventory and Analysis Data, 1976 to 2001: A Review, GTR SRS-70*; USDA Forest Service: Asheville, NC, USA, 2003.

23. Rudis, V.A. A knowledge base for FIA data uses. In *Proceedings of the Fifth Annual Forest Inventory and Analysis Symposium, GTR-WO-69, New Orleans, LA, USA, 18–20 November 2003*; McRoberts, R.E., Reams, G.A., Van Deusen, P.C., McWilliams, W., Eds.; U.S. Department of Agriculture, Forest Service: Washington, DC, USA, 2005; pp. 209–214.

24. Tinkham, W.T.; Mahoney, P.R.; Hudak, A.T.; Domke, G.M.; Falkowski, M.J.; Woodall, C.W.; Smith, A.M.S. Applications of the United States Forest Inventory and Analysis dataset: A review and future directions. *Can. J. For. Res.* **2018**, *48*, 1–18. [CrossRef]

25. Hoover, C.M.; Bush, R.; Palmer, M.; Treasure, E. Using forest inventory and analysis data to support national forest management: Regional case studies. *J. For.* **2020**, *118*, 313–323. [CrossRef]

26. Wurtzebach, Z.; DeRose, R.J.; Bush, R.R.; Goeking, S.A.; Healey, S.; Menlove, J.; Pelz, K.A.; Schultz, C.; Shaw, J.D.; Witt, C. Supporting national forest system planning with forest inventory and analysis data. *J. For.* **2020**, *118*, 289–306. [CrossRef]

27. Dugan, A.J.; Birdsey, R.; Healey, S.P.; Pan, Y.; Zhang, F.; Mo, G.; Chen, J.; Woodall, C.W.; Hernandez, A.J.; McCullough, K.; et al. Forest sector carbon analyses support land management planning and projects: Assessing the influence of anthropogenic and natural factors. *Clim. Chang.* **2017**, *144*, 207–220. [CrossRef]

28. Birdsey, R.A.; Dugan, A.J.; Healey, S.P.; Dante-Wood, K.; Zhang, F.; Mo, G.; Chen, J.M.; Hernandez, A.J.; Raymond, C.L.; McCarter, J. *Assessment of the Influence of Disturbance, Management Activities, and Environmental Factors on Carbon Stocks of U.S. National Forests, RMRS-GTR-402*; U.S. Department of Agriculture, Forest Service: Fort Collins, CO, USA, 2019.

29. Randolph, K.; Morin, R.S.; Dooley, K.; Nelson, M.D.; Jovan, S.; Woodall, C.W.; Schulz, B.K.; Perry, C.H.; Kurtz, C.M.; Oswalt, S.N.; et al. *Forest Ecosystem Health Indicators, FS-1151*; U.S. Department of Agriculture, Forest Service: Washington, DC, USA, 2020.
30. Vogt, J.T.; Koch, F.H. The evolving role of forest inventory and analysis data in invasive insect research. *Am. Entomol.* **2016**, *62*, 46–58. [CrossRef]
31. Shaw, J.D. Introduction to the special section on forest inventory and analysis. *J. For.* **2017**, *115*, 246–248. [CrossRef]
32. Dawid, A.P. Statistical inference 1. In *Encyclopedia of Statistical Sciences*; Kotz, S., Johnson, N.L., Eds.; Wiley and Sons: New York, NY, USA, 1983; Volume 4, pp. 89–105.
33. McRoberts, R.E. Probability and model-based approaches to inference for proportion forest using satellite imagery as ancillary data. *Remote Sens. Environ.* **2010**, *114*, 1017–1025. [CrossRef]
34. Cochran, W. *Sampling Techniques*; John Wiley and Sons: New York, NY, USA, 1977.
35. Neyman, J.; Jeffreys, H. Outline of a theory of statistical estimation based on the classical theory of probability. *Philos. Trans. R. Soc. Lond. Ser. A* **1937**, *236*, 333–380. [CrossRef]
36. Hansen, M.; Wendt, D. Using classified Landsat Thematic Mapper data for stratification in a statewide forest inventory. In *Proceedings of the First Annual Forest Inventory and Analysis Symposium; San Antonio, TX, USA, 2–3 November 1999*; McRoberts, R., Reams, G., Van Deusen, P., Eds.; U.S. Department of Agriculture, Forest Service, Northern Research Station: St. Paul, MN, USA, 2000; pp. 20–27.
37. McRoberts, R. Using a land cover classification based on satellite imagery to improve the precision of forest inventory area estimates. *Remote Sens. Environ.* **2002**, *81*, 36–44. [CrossRef]
38. Westfall, J.A.; Patterson, P.L.; Coulston, J.W. Post-stratified estimation: Within-strata and total sample size recommendations. *Can. J. For. Res.* **2011**, *41*, 1130–1139. [CrossRef]
39. Scott, C.; Bechtold, W.; Reams, G.; Smith, W.; Westfall, J.; Hansen, M.; Moisen, G. The enhanced forest inventory and analysis program—National sampling design and estimation procedures, GTR SRS-80. In *Sample-Based Estimators Used by the Forest Inventory and Analysis National Information Management System*; USDA Forest Service: Asheville, NC, USA, 2005; pp. 53–77.
40. McRoberts, R.; Walters, B. Statistical inference for remote sensing-based estimates of net deforestation. *Remote Sens. Environ.* **2012**, *124*, 394–401. [CrossRef]
41. McRoberts, R.E.; Liknes, G.C.; Domke, G.M. Using a remote sensing-based, percent tree cover map to enhance forest inventory estimation. *For. Ecol. Manag.* **2014**, *331*, 12–18. [CrossRef]
42. McRoberts, R.E.; Chen, Q.; Walters, B.F. Multivariate inference for forest inventories using auxiliary airborne laser scanning data. *For. Ecol. Manag.* **2017**, *401*, 295–303. [CrossRef]
43. Magnussen, S.; Andersen, H.E.; Mundhenk, P. A second look at endogenous poststratification. *For. Sci.* **2015**, *61*, 624–634. [CrossRef]
44. Bethlehem, J.G.; Keller, W.J. Linear weighting of sample survey data. *J. Off. Stat.* **1987**, *3*, 141–153.
45. Breidt, J.; Opsomer, J. Endogenous post-stratification in surveys: Classifying with a sample-fitted model. *Ann. Stat.* **2008**, *36*, 403–427. [CrossRef]
46. Stehman, S.V. Model-assisted estimation as a unifying framework for estimating the area of land cover and land-cover change from remote sensing. *Remote Sens. Environ.* **2009**, *113*, 2455–2462. [CrossRef]
47. McRoberts, R.E.; Næsset, E.; Gobakken, T.; Bollandsås, O.M.; Malimbwi, R.E.; Morgan, P.; Lundmark, T.; Wallentin, C.; Brienen, R.J.W. Inference for lidar-assisted estimation of forest growing stock volume. *Remote Sens. Environ.* **2013**, *128*, 268–275. [CrossRef]
48. McRoberts, R.E.; Næsset, E.; Gobakken, T. Estimation for inaccessible and non-sampled forest areas using model-based inference and remotely sensed auxiliary information. *Remote Sens. Environ.* **2014**, *154*, 226–233. [CrossRef]
49. McRoberts, R.E.; Næsset, E.; Gobakken, T.; Chirici, G.; Condés, S.; Hou, Z.; Saarela, S.; Chen, Q.; Ståhl, G.; Walters, B.F. Assessing components of the model-based mean square error estimator for remote sensing assisted forest applications. *Can. J. For. Res.* **2018**, *48*, 642–649. [CrossRef]
50. Rao, J.N.K.; Molina, I. *Small Area Estimation*, 2nd ed.; John Wiley and Sons Inc: Hoboken, NJ, USA, 2015; p. 480.
51. Moisen, G.; Blackard, J.; Finco, M. Small area estimation in forests affected by wildfire in the Interior West. In Proceedings of the Tenth Forest Service Remote Sensing Applications Center Conference, Salt Lake City, UT, USA, 5–9 April 2004.

52. LeMay, V.; Temesgen, H. Comparison of nearest neighbor methods for estimating basal area and stems per hectare using aerial auxiliary variables. *For. Sci.* **2005**, *51*, 109–119. [CrossRef]

53. Goerndt, M.E.; Monleon, V.J.; Temesgen, H. A comparison of small-area estimation techniques to estimate selected stand attributes using LiDAR-derived auxiliary variables. *Can. J. For. Res.* **2011**, *41*, 1189–1201. [CrossRef]

54. Mauro, F.; Monleon, V.J.; Temesgen, H.; Ford, K.R. Analysis of area level and unit level models for small area estimation in forest inventories assisted with LiDAR auxiliary information. *PLoS ONE* **2017**, *12*. [CrossRef]

55. Goerndt, M.E.; Wilson, B.T.; Aguilar, F.X. Comparison of small area estimation methods applied to biopower feedstock supply in The Northern, U.S. region. *Biomass Bioenergy* **2019**, *121*, 64–77. [CrossRef]

56. Fattorini, L. Design-based or model-based inference? The role of hybrid approaches in environmental surveys. In *Studies in Honor of Claudio Scala*; Fattorini, L., Ed.; University of Siena: Siena, Italy, 2012; pp. 173–214.

57. Corona, P.; Fattorini, L.; Franceschi, S.; Scrinzi, G.; Torresan, C. Estimation of standing wood volume in forest compartments by exploiting airborne laser scanning information: Model-based, design-based, and hybrid perspectives. *Can. J. For. Res.* **2014**, *44*, 1303–1311. [CrossRef]

58. McRoberts, R.E.; Chen, Q.; Domke, G.M.; Ståhl, G.; Saarela, S.; Westfall, J.A. Hybrid estimators for mean aboveground carbon per unit area. *For. Ecol. Manag.* **2016**, *378*, 44–56. [CrossRef]

59. Ståhl, G.; Saarela, S.; Schnell, S.; Holm, S.; Breidenbach, J.; Healey, S.P.; Patterson, P.L.; Magnussen, S.; Næsset, E.; McRoberts, R.E.; et al. Use of models in large-area forest surveys: Comparing model-assisted, model-based and hybrid estimation. *For. Ecosyst.* **2016**, *3*, 1–11. [CrossRef]

60. Saarela, S.; Schnell, S.; Grafström, A.; Tuominen, S.; Nordkvist, K.; Hyyppä, J.; Kangas, A.; Ståhl, G. Effects of sample size and model form on the accuracy of model-based estimators of growing stock volume. *Can. J. For. Res.* **2015**, *45*, 1524–1534. [CrossRef]

61. Hendee, C.W. *Forest Survey Handbook (FSH 4813.1)*; U.S. Department of Agriculture, Forest Service: Washington, DC, USA, 1967.

62. Brooks, E.; Coulston, J.; Wynne, R.; Thomas, V. Improving the precision of dynamic forest parameter estimates using Landsat. *Remote Sens. Environ.* **2016**, *179*, 162–169. [CrossRef]

63. Peterson, D.J.; Resetar, S.; Brower, J.; Diver, R. *Forest Monitoring and Remote Sensing: A Survey of Accomplishments and Opportunities for the Future, MR-1111.0-OSTP*; Rand Corporation: Washington, DC, USA, 1999.

64. Kehl, M.; Lister, A.; Scott, C.T.; Baldauf, T.; Plugge, D. Implications of sampling design and sample size for national carbon accounting systems. *Carbon Balance Manag.* **2011**, *6*, 10. [CrossRef]

65. Liknes, G.C.; Nelson, M.D.; Kaisershot, D.J. *Net Change in Forest Density, 1873-2001. Using Historical Maps to Monitor Long-Term Forest Trends, RMAP-NRS-4*; U.S. Department of Agriculture, Forest Service, Northern Research Station: Newtown Square, PA, USA, 2013.

66. Labau, V.J.; Bones, J.T.; Kingsley, N.P.; Lund, H.G.; Smith, W.B. *A History of the Forest Survey in the United States: 1830–2004, FS-877*; U.S. Department of Agriculture, Forest Service: Washington, DC, USA, 2007.

67. Bickford, C.A. The sampling design used in the forest survey of The Northeast. *J. For.* **1952**, *50*, 290–293. [CrossRef]

68. Bickford, A.; Mayer, C.; Ware, K. An efficient sampling design for forest inventory: The Northeastern forest resurvey. *J. For.* **1963**, 826–833.

69. Reams, G.; Smith, W.; Hansen, M.; Bechtold, W.; Roesch, F.; Moisen, G. The enhanced forest inventory and analysis program—National sampling design and estimation procedures. In *The Forest Inventory and Analysis Sampling Frame, GTR SRS-80*; U.S. Department of Agriculture, Forest Service: Asheville, NC, USA, 2005; pp. 21–36.

70. Riemann, R.; Tillman, K. *FIA Photointerpretation in Southern New England: A Tool to Determine Forest Fragmentation and Proximity to Human Development. NE-709*; U.S. Department of Agriculture, Forest Service: Radnor, PA, USA, 1999; pp. 1–12.

71. Rufe, P.P. *Digital Orthoimagery Base Specification V1.0, Rep 11-B5*; U.S. Geologic Survey: Reston, VA, USA, 2014.

72. Davis, D. National Aerial Imagery Program Imagery (NAIP) Information Sheet 2017. Available online: https://www.fsa.usda.gov/Assets/USDA-FSA-Public/usdafiles/APFO/support-documents/pdfs/naip_infosheet_2017.pdf (accessed on 13 December 2020).

73. Reams, G.A.; Van Deusen, P.C. The southern annual forest inventory system. *J. Agric. Biol. Environ. Stat.* **1999**, *4*, 346–360. [CrossRef]

74. Wynne, R.; Oderwald, R.; Reams, G.; Scrivani, J. Optical remote sensing for forest area estimation. *J. For.* **2000**, *98*, 31–36.

75. Teuber, K. Use of AVHRR imagery for large-scale forest inventories. *For. Ecol. Manag.* **1990**, *33–34*, 621–631. [CrossRef]

76. DeFries, R.S.; Townshend, J.R. NDVI-derived land cover classifications at a global scale. *Int. J. Remote Sens.* **1994**, *15*, 3567–3586. [CrossRef]

77. Gallo, K.P.; Eidenshink, J.C. Differences in visible and near-IR responses, and derived vegetation indices, for the NOAA-9 and NOAA-10 AVHRRs: A case study. *Photogramm. Eng. Remote Sensing* **1988**, *54*, 485–490.

78. Clark, C.A.; Cate, R.B.; Trenchard, M.H.; Boatright, J.A.; Bizzell, R.M. Mapping and classifying large ecological units. *BioScience* **1986**, *36*, 476–478. [CrossRef]

79. Zhu, Z. *Advanced Very High-Resolution Radiometer Data to Update Forest Area Estimates for Midsouth States, SO-270*; U.S. Department of Agriculture, Forest Service: New Orleans, LA, USA, 1992.

80. Roesch, F.A.; Van Deusen, P.C.; Zhu, Z. A comparison of various estimators for updating forest area coverage using AVHRR and forest inventory data. *Photogramm. Eng. Remote Sens.* **1995**, *61*, 307–311.

81. Moisen, G.G.; Edwards, T.C., Jr. Use of generalized linear models and digital data in a forest inventory of Northern Utah. *J. Agric. Biol. Environ. Stat.* **1999**, *4*, 372–390. [CrossRef]

82. Zhu, Z.; Evans, D.U.S. forest types and predicted percent forest cover from AVHRR data. *Photogramm. Eng. Remote Sens.* **1994**, *60*, 525–531.

83. Cooke, W. *Development of a Methodology for Predicting Forest Area for Large-Area Resource Monitoring, SRS-24*; U.S. Department of Agriculture, Forest Service, Southern Research Station: Starkville, MS, USA, 2001.

84. Faundeen, J.L.; Williams, D.L.; Greenhagen, C.A. Landsat yesterday and today: An American vision and an old challenge. *J. Map Geogr. Libr.* **2004**, *1*, 59–73. [CrossRef]

85. Lister, A.J. Advances in monitoring forest growth and health, Chapter 10. In *Achieving Sustainable Management of Boreal and Temperate Forests*; Stanturf, J., Ed.; Burleigh Dodds Series in Agricultural Science Series; Burleigh Dodds Science Publishing Limited: London, UK, 2019; pp. 1–30.

86. Cooke, W.; Hartsell, A. Wall-to-wall Landsat TM classifications for Georgia in support of SAFIS using FIA plots for training and verification. In *Proceedings of the Eighth Biennial Forest Service Remote Sensing Applications Conference*; Greer, J.D., Ed.; American Society for Photogrammetry and Remote Sensing: Bethesda, MD, USA, 2001; pp. 1–11.

87. McRoberts, R.E.; Wendt, D.G.; Liknes, G.C. Stratified estimation of forest inventory variables using spatially summarized stratifications. *Silva Fenn.* **2005**, *39*, 559–571. [CrossRef]

88. McRoberts, R.E.; Holden, G.R.; Nelson, M.D.; Liknes, G.C.; Gormanson, D.D. Using satellite imagery as ancillary data for increasing the precision of estimates for the forest inventory and analysis program of the USDA Forest Service. *Can. J. For. Res.* **2005**, *35*, 2968–2980. [CrossRef]

89. Hoppus, M.; Arner, S.; Lister, A. Stratifying FIA ground plots using a 3-year old MRLC forest cover map and current TM derived variables selected by "decision tree" classification. In *Proceedings of the Second Annual Forest Inventory and Analysis Symposium, GTR-SRS-47, Salt Lake City, UT, USA, 17–18 October 2000*; Reams, G.A., McRoberts, R.E., Van Deusen, P.C., Eds.; U.S. Department of Agriculture, Forest Service, Southern Research Station: Asheville, NC, USA, 2001; pp. 19–24.

90. Moisen, G.G.; Edwards Jr., T.C.; Frescino, T.S. Expanding applications, data, and models in a forest inventory of Northern Utah, USA. In *Proceedings of the North American Science Symposium, RMRS-P-12*; Aguirre-Bravo, C., Rodriguez Franco, C., Eds.; U.S. Department of Agriculture, Forest Service, Rocky Mountain Research Station: Fort Collins, CO, USA, 1999; pp. 212–218.

91. Dunham, P.; Weyermann, D.; Azuma, D. A comparison of stratification effectiveness between the National Land Cover Data set and photointerpretation in western Oregon. In *Proceedings of the Third Annual Forest Inventory and Analysis Symposium, GTR NC-230, Traverse City, MI, USA, 17–19 October 2001*; McRoberts, R.E., Reams, G.A., Van Deusen, P.C., Moser, J.W., Eds.; U.S. Department of Agriculture, Forest Service, North Central Research Station: St. Paul, MN, USA, 2002; pp. 17–19.

92. Cooke, W.; Jacobs, D. Rapid classification of Landsat TM imagery for phase 1 stratification using the Automated NDVI Threshold Supervised Classification (ANTSC) methodology. In *Proceedings of the Fourth Annual Forest Inventory and Analysis Symposium, GTR-NC-252, New Orleans, LA, USA, 19–21 November 2002*; McRoberts, R.E., Reams, G.A., Van Deusen, C.P., McWilliams, W., Cieszewski, C., Eds.; U.S. Department of Agriculture, Forest Service, North Central Research Station: St. Paul, MN, USA, 2005; pp. 81–86.

93. Homer, C.; Dewitz, J.; Fry, J.; Coan, M.; Hossain, M.; Larson, C.; Herold, N.; McKerrow, A.; VanDriel, J.N.; Wickham, J. Completion of the 2001 national land cover database for the conterminous United States. *Photogramm. Eng. Remote Sens.* **2007**, *73*, 337–341.

94. Vogelmann, J.E.; Howard, S.M.; Yang, L.; Larson, C.R.; Wylie, B.K.; Van Driel, N. Completion of the 1990s National Land Cover Data set for the conterminous United States from Landsat thematic mapper data and ancillary data sources. *Photogramm. Eng. Remote Sens.* **2001**, *67*, 650–662.

95. Huang, C.; Yang, L.; Homer, C.; Coan, M.; Rykhus, R.; Zhang, Z.; Wylie, B.; Hegge, K.; Zhu, Z.; Lister, A.; et al. Synergistic use of FIA plot data and Landsat 7 ETM+ images for large area forest mapping. In *Proceedings of the Third Annual Forest Inventory and Analysis Symposium*; McRoberts, R., Reams, G., Van Deusen, P., Moser, J., Eds.; U.S. Department of Agriculture, Forest Service, North Central Research Station: St. Paul, MN, USA, 2002; pp. 50–55.

96. Liknes, G.; Nelson, M.; Gormanson, D.; Hansen, M. The utility of the cropland data layer for forest inventory and analysis. In *Proceedings of the Third Annual Forest Inventory and Analysis Symposium, GTR NC-230, Traverse City, MI, USA, 17–19 October 2001*; McRoberts, R.E., Reams, G.A., Van Deusen, P.C., Moser, J.W., Eds.; U.S. Department of Agriculture, Forest Service, North Central Research Station: St. Paul, MN, USA, 2002; pp. 259–264.

97. Eidenshink, J.C.; Schwind, B.; Brewer, K.; Zhu, Z.-L.; Quayle, B.; Howard, S.M. A project for monitoring trends in burn severity. *Fire Ecol.* **2007**, *3*, 3–21. [CrossRef]

98. Rollins, M.; Frame, C. *The LANDFIRE Prototype Project: Nationally Consistent and Locally Relevant Geospatial Data for Wildland Fire Management, RMRS-GTR-175*; U.S. Department of Agriculture, Forest Service, Rocky Mountain Research Station: Fort Collins, CO, USA, 2006.

99. Ryan, K.; Opperman, T. LANDFIRE—A national vegetation/fuels data base for use in fuels treatment, restoration, and suppression planning. *For. Ecol. Manag.* **2013**, *294*, 208–216. [CrossRef]

100. Frescino, T.; Rollins, M. Mapping potential vegetation type for the LANDFIRE Prototype Project. In *The LANDFIRE Prototype Project: Nationally Consistent and Locally Relevant Geospatial Data for Wildland Fire Management*; USDA Forest Service, Rocky Mountain Research Station: Fort Collins, CO, USA, 2006; pp. 181–196.

101. Toney, C.; Rollins, M.; Short, K.; Frescino, T.; Tymcio, R.; Peterson, B. Use of FIA plot data in the LANDFIRE Project. In *Proceedings of the Seventh Annual Forest Inventory and Analysis Symposium, GTR WO-77, Portland, ME, USA, 3–6 October 2005*; McRoberts, R.E., Reams, G.A., Van Deusen, P.C., McWilliams, W., Eds.; U.S. Department of Agriculture, Forest Service, Washington Office: Washington, DC, USA, 2007; pp. 309–319.

102. Whittier, T.R.; Gray, A.N. Tree mortality-based fire severity classification for forest inventories: A Pacific Northwest national forests example. *For. Ecol. Manag.* **2016**, *359*, 199–209. [CrossRef]

103. Shaw, J.D.; Goeking, S.A.; Menlove, J.; Werstak, C.E. Assessment of fire effects based on forest inventory and analysis data and a long-term fire mapping data set. *J. For.* **2017**, *115*, 258–269. [CrossRef]

104. Justice, C.O.; Vermote, E.; Townshend, J.R.G.; Defries, R.; Roy, D.P.; Hall, D.K.; Salomonson, V.V.; Privette, J.L.; Riggs, G.; Strahler, A.; et al. The Moderate Resolution Imaging Spectroradiometer (MODIS): Land remote sensing for global change research. *IEEE Trans. Geosci. Remote Sens.* **1998**, *36*, 1228–1249. [CrossRef]

105. White, M.; Shaw, J.; Ramsey, R. Accuracy assessment of the vegetation continuous field tree cover product using 3954 ground plots in the south-western USA. *Int. J. Remote Sens.* **2005**, *26*, 2699–2704. [CrossRef]

106. Nelson, M.D.; McRoberts, R.E.; Hansen, M.C. Forest land area estimates from Vegetation Continuous Fields. In *Proceedings of the 10th Forest Service Remote Sensing Applications Conference*; American Society for Photogrammetry and Remote Sensing: Salt Lake City, UT, USA, 2004; pp. 1–6.

107. Nelson, M.D.; McRoberts, R.E.; Holden, G.R.; Bauer, M.E. Effects of satellite image spatial aggregation and resolution on estimates of forest land area. *Int. J. Remote Sens.* **2009**, *30*, 1913–1940. [CrossRef]

108. Liknes, G.C.; Nelson, M.D.; McRoberts, R.E. Evaluating classified MODIS satellite imagery as a stratification tool. In Proceedings of the Joint Meeting of the 6th International Symposium on Spatial Accuracy Assessment in Natural Resources and Environmental Sciences and the 15th Annual Conference of the International Environmetrics Society, Portland, ME, USA, 28 June–1 July 2004.

109. Holden, G.; Nelson, M.; McRoberts, R. Accuracy assessment of FIA's nationwide biomass mapping products: Results from the north central FIA region. In *Proceedings of the Fifth Annual Forest Inventory and Analysis Symposium, GTR-WO-69, New Orleans, LA, USA, 18–20 November 2003*; McRoberts, R.E., Reams, G.A., Van Deusen, P.C., McWilliams, W., Eds.; USDA Forest Service, Washington Office: Washington, DC, USA, 2005; pp. 139–147.

110. Goeking, S.A.; Patterson, P.L. *Stratifying to Reduce Bias Caused by High Nonresponse Rates: A Case Study from New Mexico's Forest Inventory, RMRS-RN-59*; USDA Forest Service, Rocky Mountain Research Station: Fort Collins, CO, USA, 2013.

111. Witten, I.H.; Frank, E.; Hall, M.A.; Pal, C.J. *Data Mining: Practical Machine Learning Tools and Techniques*; Morgan Kaufmann: New York, NY, USA, 2016; ISBN 0-12-804357-1.

112. Moisen, G.G.; Frescino, T.S. Comparing five modelling techniques for predicting forest characteristics. *Ecol. Modell.* **2002**, *157*, 209–225. [CrossRef]

113. McRoberts, R.E.; Nelson, M.D.; Wendt, D.G. Stratified estimation of forest area using satellite imagery, inventory data, and the k-Nearest neighbor technique. *Remote Sens. Environ.* **2002**, *82*, 457–468. [CrossRef]

114. Lister, A.; Hoppus, M.; Czaplewski, R.L. K-nearest neighbor imputation of forest inventory variables in New Hampshire. In *Proceedings of the Tenth Forest Service Remote Sensing Applications Center Conference, Salt Lake City, UT, USA, 5–9 April 2004*; Greer, J., Ed.; ASPRS: Bethesda, MD, USA.

115. Ohmann, J.L.; Gregory, M.J. Predictive mapping of forest composition and structure with direct gradient analysis and nearest- neighbor imputation in coastal Oregon, U.S.A. *Can. J. For. Res.* **2002**, *32*, 725–741. [CrossRef]

116. Nelson, M.D.; McRoberts, R.E.; Liknes, G.C.; Holden, G.R. Comparing forest/nonforest classifications of Landsat TM imagery for stratifying FIA estimates of forest land area. In *Proceedings of the Fourth Annual Forest Inventory and Analysis Symposium, GTR-NC-252, New Orleans, LA, USA, 19–21 November 2002*; McRoberts, R.E., Reams, G.A., Van Deusen, C.P., McWilliams, W., Cieszewski, C., Eds.; U.S. Department of Agriculture, Forest Service, North Central Research Station: St. Paul, MN, USA, 2005; pp. 121–128.

117. Bivand, R.S.; Pebesma, E.; Gómez-Rubio, V. *Applied Spatial Data Analysis with R*; Springer: New York, NY, USA, 2013.

118. Rulequest Research. *Cubist and See5 Data Mining Software*; Rulequest Research: Empire Bay, NSW, Australia, 2020.

119. Ruefenacht, B.; Liknes, G.; Lister, A.J.; Fisk, H.; Wendt, D. Evaluation of open source data mining software packages. In Proceedings of the Forest Inventory and Analysis (FIA) Symposium 2008, RMRS-P-56CD, Park City, UT, USA, 21–23 October 2008.

120. Blackard, J.A.; Finco, M.V.; Helmer, E.H.; Holden, G.R.; Hoppus, M.L.; Jacobs, D.M.; Lister, A.J.; Moisen, G.G.; Nelson, M.D.; Riemann, R.; et al. Mapping, U.S. forest biomass using nationwide forest inventory data and moderate resolution information. *Remote Sens. Environ.* **2008**, *112*, 1658–1677. [CrossRef]

121. Ruefenacht, B.; Finco, M.; Nelson, M.; Czaplewski, R.; Helmer, E.; Blackard, J.A.; Holden, G.; Lister, A.; Salajanu, D.; Weyermann, D.; et al. Conterminous, U.S. and Alaska forest type mapping using forest inventory and analysis data. *Photogramm. Eng. Remote Sens.* **2008**, *74*, 1379–1388. [CrossRef]

122. Wilson, B.T.; Lister, A.J.; Riemann, R.I. A nearest-neighbor imputation approach to mapping tree species over large areas using forest inventory plots and moderate resolution raster data. *For. Ecol. Manag.* **2012**, *271*, 182–198. [CrossRef]

123. Wilson, B.T.; Woodall, C.W.; Griffith, D.M. Imputing forest carbon stock estimates from inventory plots to a nationally continuous coverage. *Carbon Balance Manag.* **2013**, *8*, 1. [CrossRef]

124. Ellenwood, J.R.; Krist, F.J.; Romero, S.A. *National Individual Tree Species Atlas, FHTET 15-01*; United States Forest Service, Forest Health Protection, Forest Health Technology Enterprise Team: Fort Collins, CO, USA, 2015.

125. FHTET. *U.S. Forest Service National Insect and Disease Risk Map*; United States Forest Service, Forest Health Protection, Forest Health Technology Enterprise Team: Fort Collins, CO, USA, 2015.

126. Woodcock, C.E.; Allen, R.; Anderson, M.; Belward, A.; Bindschadler, R.; Cohen, W.; Gao, F.; Goward, S.N.; Helder, D.; Helmer, E.; et al. Free access to Landsat imagery. *Science* **2008**, *320*, 1011. [CrossRef]

127. Wulder, M.A.; Masek, J.G.; Cohen, W.B.; Loveland, T.R.; Woodcock, C.E. Opening the archive: How free data has enabled the science and monitoring promise of Landsat. *Remote Sens. Environ.* **2012**, *122*, 2–10. [CrossRef]

128. USGS Earth Resources Observation and Science (EROS) Center Data Discovery, 2020. Available online: https://www.usgs.gov/centers/eros (accessed on 13 December 2020).

129. European Union. EU Copernicus—The EU Earth Observation and Monitoring Program. Fact Sheet, 2020. Available online: https://www.copernicus.eu/sites/default/files/2019-06/The_EU_Earth_Observation_and_Monitoring_Programme-EN-20190405-WEB.pdf (accessed on 13 December 2020).

130. Roy, D.P.; Ju, J.; Kline, K.; Scaramuzza, P.L.; Kovalskyy, V.; Hansen, M.; Loveland, T.R.; Vermote, E.; Zhang, C. Web-enabled Landsat Data (WELD): Landsat ETM+ composited mosaics of the conterminous United States. *Remote Sens. Environ.* **2010**, *114*, 35–49. [CrossRef]

131. Masek, J.G.; Vermote, E.F.; Saleous, N.E.; Wolfe, R.; Hall, F.G.; Huemmrich, K.F.; Gao, F.; Kutler, J.; Teng-Kui, L. A Landsat surface reflectance data set for North America, 1990–2000. *Geosci. Remote Sens. Lett.* **2006**, *3*, 68–72. [CrossRef]

132. Young, N.E.; Anderson, R.S.; Chignell, S.M.; Vorster, A.G.; Lawrence, R.; Evangelista, P.H. A survival guide to Landsat preprocessing. *Ecology* **2017**, *98*, 920–932. [CrossRef] [PubMed]

133. Banskota, A.; Kayastha, N.; Falkowski, M.; Wulder, M.; Froese, R.; White, J. Forest monitoring using Landsat time series data: A review. *Can. J. Remote Sens.* **2014**, *40*, 362–384. [CrossRef]

134. Huang, C.; Goward, S.N.; Masek, J.G.; Thomas, N.; Zhu, Z.; Vogelmann, J.E. An automated approach for reconstructing recent forest disturbance history using dense Landsat time series stacks. *Remote Sens. Environ.* **2010**, *114*, 183–198. [CrossRef]

135. Goward, S.N.; Masek, J.G.; Cohen, W.; Moisen, G.; Collatz, G.J.; Healey, S.; Houghton, R.; Huang, C.; Kennedy, R.; Law, B. Forest disturbance and North American carbon flux. *Eos Trans. Am. Geophys. Union* **2008**, *89*, 105–106. [CrossRef]

136. Goward, S.N.; Huang, C.; Zhao, F.; Schleeweis, K.G.; Rishmawi, K.; Lindsey, M.; Dungan, J.L.; Michaelis, A.R. *Dataset: NACP NAFD Project: Forest Disturbance History from Landsat, 1986–2010*; Oak Ridge National Laboratory DAAC: Oak Ridge, TN, USA, 2015.

137. Masek, J.G.; Goward, S.N.; Kennedy, R.E.; Cohen, W.B.; Moisen, G.G.; Schleeweis, K.; Huang, C. United States forest disturbance trends observed using Landsat time series. *Ecosystems* **2013**, *16*, 1087–1104. [CrossRef]

138. Schleeweis, K.; Goward, S.N.; Huang, C.; Dwyer, J.L.; Dungan, J.L.; Lindsey, M.A.; Michaelis, A.; Rishmawi, K.; Masek, J.G. Selection and quality assessment of Landsat data for the North American forest dynamics forest history maps of the US. *Int. J. Digit. Earth* **2016**, *9*, 963–980. [CrossRef]

139. Schleeweis, K.G.; Moisen, G.G.; Schroeder, T.A.; Toney, C.; Freeman, E.A.; Goward, S.N.; Huang, C.; Dungan, J.L. U.S. national maps attributing forest change: 1986–2010. *Forests* **2020**, *11*, 653. [CrossRef]

140. Zhao, F.; Huang, C.; Goward, S.N.; Schleeweis, K.; Rishmawi, K.; Lindsey, M.A.; Denning, E.; Keddell, L.; Cohen, W.B.; Yang, Z.; et al. Development of Landsat-based annual U.S. forest disturbance history maps (1986–2010) in support of the North American Carbon Program (NACP). *Remote Sens. Environ.* **2018**, *209*, 312–326. [CrossRef]

141. Huang, C.; Ling, P.-Y.; Zhu, Z. North Carolina's forest disturbance and timber production assessed using time series Landsat observations. *Int. J. Digit. Earth* **2015**, *8*, 947–969. [CrossRef]

142. Tao, X.; Huang, C.; Zhao, F.; Schleeweis, K.; Masek, J.; Liang, S. Mapping forest disturbance intensity in North and South Carolina using annual Landsat observations and field inventory data. *Remote Sens. Environ.* **2019**, *221*, 351–362. [CrossRef]

143. Schroeder, T.A.; Schleeweis, K.G.; Moisen, G.G.; Toney, C.; Cohen, W.B.; Freeman, E.A.; Yang, Z.; Huang, C. Testing a Landsat-based approach for mapping disturbance causality in U.S. forests. *Remote Sens. Environ.* **2017**, *195*, 230–243. [CrossRef]

144. Huang, C.; Goward, S.N.; Masek, J.G.; Gao, F.; Vermote, E.F.; Thomas, N.; Schleeweis, K.; Kennedy, R.E.; Zhu, Z.; Eidenshink, J.C.; et al. Development of time series stacks of Landsat images for reconstructing forest disturbance history. *Int. J. Digit. Earth* **2009**, *2*, 195–218. [CrossRef]

145. Stueve, K.; Housman, I.; Zimmerman, P.; Nelson, M.; Webb, J.; Perry, C.; Chastain, R.; Gormanson, D.; Huang, C.; Healey, S.; et al. Snow-covered Landsat time series stacks improve automated disturbance mapping accuracy in forested landscapes. *Remote Sens. Environ.* **2011**, *115*, 3203–3219. [CrossRef]

146. Garner, J.D.; Nelson, M.D.; Tavernia, B.G.; Perry, C.H.H.; Housman, I.W. Mapping forest canopy disturbance in the Upper Great Lakes, USA. In *Proceedings of the 2015 Forest Inventory and Analysis (FIA) symposium, PNW-GTR-931, Portland, OR, USA, 8–10 December 2015*; Stanton, S., Christensen, G.A., Eds.; Department of Agriculture, Forest Service, Pacific Northwest Research Station: Portland, OR, USA, 2015; pp. 363–367.

147. Garner, J.D.; Nelson, M.D.; Tavernia, B.G.; Housman, I.W.; Perry, C.H. *Early Successional Forest and Land Cover Geospatial Dataset of the Upper Midwest, RDS-2016-0001*; USDA Forest Service, Research Data Archive: Fort Collins, CO, USA, 2016.

148. Tavernia, B.G.; Nelson, M.D.; Garner, J.D.; Perry, C.H. Spatial characteristics of early successional habitat across the upper great lakes states. *For. Ecol. Manag.* **2016**, *372*, 164–174. [CrossRef]

149. Nelson, M.; Stueve, K.; Perry, C.; Gormanson, D.; Huang, C.; Healey, S. *Mapping Young Forest in Wisconsin, NRS-RN-154*; U.S. Department of Agriculture, Forest Service, Northern Research Station: Newtown Square, PA, USA, 2012.

150. Powell, S.; Cohen, W.; Healey, S.; Kennedy, R.; Moisen, G.; Pierce, K.; Ohmann, J. Quantification of live aboveground forest biomass dynamics with Landsat time-series and field inventory data: A comparison of empirical modeling approaches. *Remote Sens. Environ.* **2010**, *114*, 1053–1068. [CrossRef]

151. Brown, J.; Lister, A.J.; Fajvan, M.A.; Ruefenacht, B.; Mazzarella, C. Modeling forest ecosystem changes resulting from surface coal mining in West Virginia. In *Proceedings of the 2010 Joint Meeting of the Forest Inventory and Analysis (FIA) Symposium and the Southern Mensurationists, GTR SRS-157, Knoxville, TN, USA, 5–7 October 2010*; McWilliams, W., Roesch, F.A., Eds.; U.S. Department of Agriculture, Forest Service, Southern Research Station: Asheville, NC, USA, 2012; pp. 67–75.

152. Schroeder, T.A.; Healey, S.P.; Moisen, G.G.; Frescino, T.S.; Cohen, W.B.; Huang, C.; Kennedy, R.E.; Yang, Z. Improving estimates of forest disturbance by combining observations from Landsat time series with U.S. Forest Service Forest Inventory and Analysis data. *Remote Sens. Environ.* **2014**, *154*, 61–73. [CrossRef]

153. Moisen, G.G.; Meyer, M.C.; Schroeder, T.A.; Liao, X.; Schleeweis, K.G.; Freeman, E.A.; Toney, C. Shape selection in Landsat time series: A tool for monitoring forest dynamics. *Glob. Chang. Biol.* **2016**, *22*, 3518–3528. [CrossRef]

154. Kennedy, R.E.; Yang, Z.; Cohen, W.B. Detecting trends in forest disturbance and recovery using yearly Landsat time series: 1. LandTrendr—Temporal segmentation algorithms. *Remote Sens. Environ.* **2010**, *114*, 2897–2910. [CrossRef]

155. Cohen, W.B.; Yang, Z.; Kennedy, R. Detecting trends in forest disturbance and recovery using yearly Landsat time series: 2. TimeSync—Tools for calibration and validation. *Remote Sens. Environ.* **2010**, *114*, 2911–2924. [CrossRef]

156. Ohmann, J.L.; Gregory, M.J.; Roberts, H.M.; Cohen, W.B.; Kennedy, R.E.; Yang, Z. Mapping change of older forest with nearest-neighbor imputation and Landsat time-series. *For. Ecol. Manag.* **2012**, *272*, 13–25. [CrossRef]

157. Bright, B.C.; Hudak, A.T.; Kennedy, R.E.; Meddens, A.J.H. Landsat time series and lidar as predictors of live and dead basal area across five bark beetle-affected forests. *IEEE J. Sel. Top. Appl. Earth Obs. Remote Sens.* **2014**, *7*, 3440–3452. [CrossRef]

158. Bell, D.M.; Cohen, W.B.; Reilly, M.; Yang, Z. Visual interpretation and time series modeling of Landsat imagery highlight drought's role in forest canopy declines. *Ecosphere* **2018**, *9*, e02195. [CrossRef]

159. Cohen, W.B.; Yang, Z.; Stehman, S.V.; Schroeder, T.A.; Bell, D.M.; Masek, J.G.; Huang, C.; Meigs, G.W. Forest disturbance across the conterminous United States from 1985–2012: The emerging dominance of forest decline. *For. Ecol. Manag.* **2016**, *360*, 242–252. [CrossRef]

160. Gray, A.N.; Cohen, W.B.; Yang, Z.; Pfaff, E. Integrating TimeSync disturbance detection and repeat forest inventory to predict carbon flux. *Forests* **2019**, *10*, 984. [CrossRef]

161. Filippelli, S.K.; Falkowski, M.J.; Hudak, A.T.; Fekety, P.A.; Vogeler, J.C.; Khalyani, A.H.; Rau, B.M.; Strand, E.K. Monitoring pinyon-juniper cover and aboveground biomass across the Great Basin. *Environ. Res. Lett.* **2020**, *15*, 025004. [CrossRef]

162. Lister, A.J.; Leites, L.P. A Sentinel satellite-based forest ecosystem change detection system. In *Proceedings of the 2017 Forest Inventory and Analysis (FIA) Science Stakeholder Meeting, RMRS-P-75, Park City, UT, USA, 24–26 October 2017*; USDA Forest Service, Rocky Mountain Research Station: Fort Collins, CO, USA, 2017; pp. 45–48.

163. Cohen, W.B.; Yang, Z.; Healey, S.P.; Kennedy, R.E.; Gorelick, N. A LandTrendr multispectral ensemble for forest disturbance detection. *Remote Sens. Environ.* **2018**, *205*, 131–140. [CrossRef]

164. Healey, S.P.; Cohen, W.B.; Yang, Z.; Kenneth Brewer, C.; Brooks, E.B.; Gorelick, N.; Hernandez, A.J.; Huang, C.; Joseph Hughes, M.; Kennedy, R.E.; et al. Mapping forest change using stacked generalization: An ensemble approach. *Remote Sens. Environ.* **2018**, *204*, 717–728. [CrossRef]

165. Wilson, B.T.; Knight, J.F.; McRoberts, R.E. Harmonic regression of Landsat time series for modeling attributes from national forest inventory data. *ISPRS J. Photogramm. Remote Sens.* **2018**, *137*, 29–46. [CrossRef]

166. Derwin, J.M.; Thomas, V.A.; Wynne, R.H.; Coulston, J.W.; Liknes, G.C.; Bender, S.; Blinn, C.E.; Brooks, E.B.; Ruefenacht, B.; Benton, R.; et al. Estimating tree canopy cover using harmonic regression coefficients derived from multitemporal Landsat data. *Int. J. Appl. Earth Obs.* **2020**, *86*, 101985. [CrossRef]

167. Gorelick, N.; Hancher, M.; Dixon, M.; Ilyushchenko, S.; Thau, D.; Moore, R. Google Earth Engine: Planetary-scale geospatial analysis for everyone. *Remote Sens. Environ.* **2017**, *202*, 18–27. [CrossRef]

168. Hansen, M.C.; Potapov, P.V.; Moore, R.; Hancher, M.; Turubanova, S.A.; Tyukavina, A.; Thau, D.; Stehman, S.V.; Goetz, S.J.; Loveland, T.R.; et al. High-resolution global maps of 21st-century forest cover change. *Science* **2013**, *342*, 850–853. [CrossRef] [PubMed]

169. Curtis, P.G.; Slay, C.M.; Harris, N.L.; Tyukavina, A.; Hansen, M.C. Classifying drivers of global forest loss. *Science* **2018**, *361*, 1108. [CrossRef] [PubMed]

170. Kennedy, R.E.; Yang, Z.; Gorelick, N.; Braaten, J.; Cavalcante, L.; Cohen, W.B.; Healey, S. Implementation of the LandTrendr algorithm on Google Earth Engine. *Remote Sens.* **2018**, *10*. [CrossRef]

171. Zhu, Z.; Woodcock, C.E. Continuous change detection and classification of land cover using all available Landsat data. *Remote Sens. Environ.* **2014**, *144*, 152–171. [CrossRef]

172. Brown, J.F.; Tollerud, H.J.; Barber, C.P.; Zhou, Q.; Dwyer, J.L.; Vogelmann, J.E.; Loveland, T.R.; Woodcock, C.E.; Stehman, S.V.; Zhu, Z.; et al. Lessons learned implementing an operational continuous United States national land change monitoring capability: The Land Change Monitoring, Assessment, and Projection (LCMAP) approach. *Remote Sens. Environ.* **2020**, *238*, 111356. [CrossRef]

173. Pengra, B.W.; Stehman, S.V.; Horton, J.A.; Dockter, D.J.; Schroeder, T.A.; Yang, Z.; Cohen, W.B.; Healey, S.P.; Loveland, T.R. Quality control and assessment of interpreter consistency of annual land cover reference data in an operational national monitoring program. *Remote Sens. Environ.* **2020**, *238*, 111261. [CrossRef]

174. Ioup, E.; Lin, B.; Sample, J.; Shaw, K.; Rabemanantsoa, A.; Reimbold, J. Geospatial web services: Bridging the gap between OGC and web services, Chapter 4. In *Geospatial Services and Applications for the Internet*; Sample, J.T., Shaw, K., Tu, S., Abdelguerfi, M., Eds.; Springer: Boston, MA, USA, 2008; pp. 73–93, ISBN 978-0-387-74673-9.

175. Goeking, S.; Moisen, G.; Megown, K.; Toombs, J. Prefield methods: Streamlining forest or nonforest determinations to increase inventory efficiency. In *Proceedings of the Eighth Annual Forest Inventory and Analysis Symposium, Monterey, CA, USA, 16–19 October 2006*; McRoberts, R.E., Reams, G.A., Van Deusen, P.C., McWilliams, W., Eds.; USDA Forest Service, Washington Office: Washington, DC, USA, 2009; pp. 351–354.

176. Torrey, C.; Liknes, G.; Lister, A.; Meneguzzo, D. Assessing alternative measures of tree canopy cover: Photo-interpreted NAIP and ground-based estimates. In *Proceedings of the 2010 Joint Meeting of the Forest Inventory and Analysis (FIA) Symposium and the Southern Mensurationists, GTR SRS-157, Knoxville, TN, USA, 5–7 October 2010*; McWilliams, W., Roesch, F.A., Eds.; U.S. Department of Agriculture, Forest Service, Southern Research Station: Asheville, NC, USA, 2012; pp. 209–215.

177. Lister, A.J.; Scott, C.T.; Rasmussen, S. Inventory methods for trees in nonforest areas in the great plains states. *Environ. Monit. Assess.* **2012**, *184*, 2465–2474. [CrossRef] [PubMed]

178. Westfall, J.A.; Lister, A.J.; Scott, C.T.; Weber, T.A. Double sampling for post-stratification in forest inventory. *Eur. J. For. Res.* **2019**, *138*, 375–382. [CrossRef]

179. Frescino, T.S.; Moisen, G.G.; Megown, K.A.; Nelson, V.J.; Freeman, E.A.; Patterson, P.L.; Finco, M.; Brewer, K.; Menlove, J. *Nevada Photo-Based Inventory Pilot (NPIP) Photo Sampling Procedures, RMRS-GTR-222*; U.S. Department of Agriculture, Forest Service, Rocky Mountain Research Station: Fort Collins, CO, USA, 2009.

180. Lister, T.W.; Lister, A.J.; Alexander, E. Land use change monitoring in Maryland using a probabilistic sample and rapid photointerpretation. *Appl. Geogr.* **2014**, *51*, 1–7. [CrossRef]

181. Lister, A.; Lister, T.; Weber, T. Semi-automated sample-based forest degradation monitoring with photointerpretation of high-resolution imagery. *Forests* **2019**, *10*, 896. [CrossRef]

182. Nowak, D.J.; Greenfield, E.J. Evaluating the National Land Cover Database tree canopy and impervious cover estimates across the conterminous United States: A comparison with photo-interpreted estimates. *Environ. Manag.* **2010**, *46*, 378–390. [CrossRef] [PubMed]

183. Nowak, D.J.; Greenfield, E.J. Tree and impervious cover in the United States. *Landsc. Urban Plan.* **2012**, *107*, 21–30. [CrossRef]

184. Nowak, D.J.; Greenfield, E.J. Declining urban and community tree cover in the United States. *Urban For. Urban Green.* **2018**, *32*, 32–55. [CrossRef]
185. Nowak, D.J.; Maco, S.; Binkley, M. i-Tree: Global tools to assess tree benefits and risks to improve forest management. *Arboric. Consult.* **2018**, *51*, 10–13.
186. Goeking, S.A.; Liknes, G.C.; Lindblom, E.; Chase, J.; Jacobs, D.M.; Benton, R. A GIS-based tool for estimating tree canopy cover on fixed-radius plots using high-resolution aerial imagery. In *Proceedings of the Forest Inventory and Analysis (FIA) Symposium 2012, Baltimore, MD, USA, 4–6 December 2012*; Morin, R.S., Liknes, G.C., Eds.; U.S. Department of Agriculture, Forest Service, Northern Research Station: Newtown Square, PA, USA, 2012; pp. 237–241.
187. Webb, J.; Brewer, C.K.; Daniels, N.; Maderia, C.; Hamilton, R.; Finco, M.; Megown, K.A.; Lister, A.J. Image-based change estimation for land cover and land use monitoring. In *Proceedings of the Forest Inventory and Analysis (FIA) Symposium 2012, Baltimore, MD, USA, 4–6 December 2012*; Morin, R.S., Liknes, G.C., Eds.; U.S. Department of Agriculture, Forest Service, Northern Research Station: Newtown Square, PA, USA, 2012; pp. 46–53.
188. National Association of State Foresters (NASF). *State Forester Strategic Vision and Priorities for the Forest Inventory and Analysis Program, NASF-2015-06 2015*; National Association of State Foresters: Washington, DC, USA, 2015.
189. Frescino, T.S.; Moisen, G.G.; DeBlander, L.; Guerin, M. The investigation of classification methods of high-resolution imagery. In *Proceedings of the Seventh Annual Forest Inventory and Analysis Symposium, GTR WO-77, Portland, ME, USA, 3–6 October 2005*; McRoberts, R.E., Reams, G.A., Van Deusen, P.C., McWilliams, W., Eds.; U.S. Department of Agriculture, Forest Service: Washington, DC, USA, 2007; pp. 161–169.
190. Meneguzzo, D.M.; Liknes, G.C.; Nelson, M.D. Mapping trees outside forests using high-resolution aerial imagery: A comparison of pixel- and object-based classification approaches. *Environ. Monit. Assess.* **2013**, *185*, 6261–6275. [CrossRef]
191. Hogland, J.S.; Anderson, N.M.; Chung, W.; Wells, L. Estimating forest characteristics using NAIP imagery and ArcObjects. In *Proceedings of the 2014 ESRI Users Conference, San Diego, CA, USA, 14-18 July 2014*; Environmental Systems Research Institute, Inc: Redlands, CA, USA, 2014; pp. 155–181.
192. Hogland, J.; Anderson, N.; St. Peter, J.; Drake, J.; Medley, P. Mapping forest characteristics at fine resolution across large landscapes of the southeastern United States using NAIP imagery and FIA field plot data. *ISPRS Int. J. Geo Inf.* **2018**, *7*, 140. [CrossRef]
193. Chang, T.; Rasmussen, B.P.; Dickson, B.G.; Zachmann, L.J. Chimera: A multi-task recurrent convolutional neural network for forest classification and structural estimation. *Remote Sens.* **2019**, *11*, 768. [CrossRef]
194. O'Neil-Dunne, J.P.M.; MacFaden, S.W.; Royar, A.R.; Pelletier, K.C. An object-based system for LiDAR data fusion and feature extraction. *Geocarto Int.* **2013**, *28*, 227–242. [CrossRef]
195. Marcello, J.; Rodríguez-Esparragón, D.; Moreno, D. Comparison of land cover maps using high resolution multispectral and hyperspectral imagery. In Proceedings of the IGARSS 2018 IEEE International Geoscience and Remote Sensing Symposium, Valencia, Spain, 22–27 July 2018; pp. 7312–7315.
196. Lister, A.J.; Kahler, H.; Clark, A.; Zumbrun, F. Imputing forest inventory data to stands formed by image segmentation in Maryland's Green Ridge State Forest. In *Proceedings of the 6th Southern Forestry and Natural Resources GIS Conference, Athens, GA, USA*; Bettinger, P., Merry, K., Fei, S., Drake, J., Nibbelink, N., Hepinstall, J., Eds.; Warnell School of Forestry and Natural Resources: Athens, GA, USA, 2008.
197. Liknes, G.C.; Perry, C.H.; Meneguzzo, D.M. Assessing tree cover in agricultural landscapes using high-resolution aerial imagery. *J. Terr. Obs.* **2010**, *2*, 38–55.
198. Riemann, R.; O'Neil-Dunne, J.; Liknes, G.C. Building capacity for providing canopy cover and canopy height at FIA plot locations using high-resolution imagery and leaf-off LiDAR. In *Proceedings of the Forest Inventory and Analysis (FIA) Symposium 2012, Baltimore, MD, USA, 4–6 December 2012*; Morin, R.S., Liknes, G.C., Eds.; U.S. Department of Agriculture, Forest Service, Northern Research Station: Newtown Square, PA, USA, 2012; pp. 242–247.
199. Riemann, R.; Liknes, G.; O'Neil-Dunne, J.; Toney, C.; Lister, T. Comparative assessment of methods for estimating tree canopy cover across a rural-to-urban gradient in the mid-Atlantic region of the USA. *Environ. Monit. Assess.* **2016**, *188*, 297. [CrossRef] [PubMed]

200. Meneguzzo, D. Remote Sensing-based Approaches for Large-scale Comprehensive Assessments of Tree Cover and Windbreaks in the Great Plains Region of the United States. Ph.D. Thesis, University of Minnesota, St. Paul, MN, USA, 2020.

201. Liknes, G.C.; Meneguzzo, D.M.; Kellerman, T.A. Shape indexes for semi-automated detection of windbreaks in thematic tree cover maps from the central United States. *Int. J. Appl. Earth Obs. Geoinf.* **2017**, *59*, 167–174. [CrossRef]

202. Paull, D.A.; Meneguzzo, D.M.; Gonzalez, R.M.; Garcia, D.L.; Marcotte, A.L.; Liknes, G.C.; Finney, T.N. *High-Resolution Urban Land Cover of Kansas (2015), RDS-2019-0052*; Forest Service Research Data Archive: Fort Collins, CO, USA, 2019.

203. Paull, D.A.; Whitson, J.W.; Marcotte, A.L.; Liknes, G.C.; Meneguzzo, D.M.; Kellerman, T.A. *High-Resolution Land Cover of Kansas (2015), RDS-2017-0025*; Forest Service Research Data Archive: Fort Collins, CO, USA, 2017.

204. Kellerman, T.A.; Meneguzzo, D.M.; Vaitkus, M.; White, M.; Ossell, R.; Sorsen, N.; Stannard, J.; Gift, T.; Cox, J.; Liknes, G.C. *High-Resolution Land Cover of Nebraska (2014), RDS-2019-0038*; Forest Service Research Data Archive: Fort Collins, CO, USA, 2019.

205. Webb, J.; Clark, A.; Schaaf, A.; Clark, J.; Moss, S.; Cox, S.; Jardine, M.; Adkins, S. *Forest Structure Estimates from NAIP Point Cloud Data*; Forest Service, Remote Sensing Applications Center: Salt Lake City, UT, USA, 2017.

206. Strunk, J.; Packalen, P.; Gould, P.; Gatziolis, D.; Maki, C.; Andersen, H.-E.; McGaughey, R.J. Large area forest yield estimation with pushbroom digital aerial photogrammetry. *Forests* **2019**, *10*. [CrossRef]

207. Gatziolis, D. Forest stand canopy structure attribute estimation from high resolution digital airborne imagery. In *Proceedings of the Monitoring Science and Technology Symposium: Unifying Knowledge for Sustainability in the Western Hemisphere, Denver, CO, USA, 20–24 September 2004*; Aguirre-Bravo, C., Pellicane, P., Burns, D., Draggan, S., Eds.; U.S. Department of Agriculture, Forest Service, Rocky Mountain Research Station: Fort Collins, CO, USA, 2006; pp. 783–789.

208. Lefsky, M.A.; Cohen, W.B.; Parker, G.G.; Harding, D.J. Lidar Remote Sensing for Ecosystem Studies: Lidar, an emerging remote sensing technology that directly measures the three-dimensional distribution of plant canopies, can accurately estimate vegetation structural attributes and should be of particular interest to forest, landscape, and global ecologists. *BioScience* **2002**, *52*, 19–30. [CrossRef]

209. Gatziolis, D. Precise FIA plot registration using field and dense LIDAR data. In *Proceedings of the Eighth Annual Forest Inventory and Analysis Symposium, Monterey, CA, USA, 16–19 October 2006*; McRoberts, R.E., Reams, G.A., Van Deusen, P.C., McWilliams, W., Eds.; U.S. Department of Agriculture, Forest Service: Washington, DC, USA, 2009; pp. 243–249.

210. Gatziolis, D. Advancements in LiDAR-based registration of FIA field plots. In *Proceedings of the Forest Inventory and Analysis (FIA) Symposium 2012, Baltimore, MD, USA, 4–6 December 2012*; Morin, R.S., Liknes, G.C., Eds.; U.S. Department of Agriculture, Forest Service, Northern Research Station: Newtown Square, PA, USA, 2012; pp. 432–437.

211. Schrader-Patton, C.; Liknes, G.C.; Gatziolis, D.; Wing, B.M.; Nelson, M.D.; Miles, P.D.; Bixby, J.; Wendt, D.G.; Kepler, D.; Schaaf, A. Refining FIA plot locations using LiDAR point clouds. In *Proceedings of the 2015 Forest Inventory and Analysis (FIA) Symposium, PNW-GTR-931, Portland, OR, USA, 8–10 December 2015*; Stanton, S., Christensen, G.A., Eds.; U.S. Department of Agriculture, Forest Service, Pacific Northwest Research: Portland, OR, USA, 2015; pp. 247–252.

212. Gatziolis, D. LIDAR-derived site index in the U.S. pacific northwest–challenges and opportunities. In *Proceedings of ISPRS Workshop on Laser Scanning 2007 and SilviLaser 2007, Espoo, Finland, 12–14 September 2007*; ISPRS Archives: Hanover, Germany, 2007; pp. 136–143.

213. Andersen, H.E.; Reutebuch, S.E.; McGaughey, R.J. A rigorous assessment of tree height measurements obtained using airborne lidar and conventional field methods. *Can. J. Remote Sens.* **2006**, *32*, 355–366. [CrossRef]

214. Li, Y. A comparison of forest height prediction from FIA field measurement and LiDAR data via spatial models. In *Proceedings of the Forest Inventory and Analysis (FIA) Symposium 2008, Park City, UT, USA, 21–23 October 2008*; McWilliams, W., Moisen, G., Czaplewski, R., Eds.; U.S. Department of Agriculture, Forest Service, Rocky Mountain Research Station: Fort Collins, CO, USA, 2009.

215. Gopalakrishnan, R.; Thomas, V.A.; Coulston, J.W.; Wynne, R.H. Prediction of canopy heights over a large region using heterogeneous lidar datasets: Efficacy and challenges. *Remote Sens.* **2015**, *7*, 11036–11060. [CrossRef]

216. Gatziolis, D.; Fried, J.S.; Monleon, V.S. Challenges to estimating tree height via LiDAR in closed-canopy forests: A parable from western Oregon. *For. Sci.* **2010**, *56*, 139–155.

217. Skowronski, N.; Clark, K.; Nelson, R.; Hom, J.; Patterson, M. Remotely sensed measurements of forest structure and fuel loads in the Pinelands of New Jersey. *Remote Sens. Environ.* **2007**, *108*, 123–129. [CrossRef]

218. Johnson, K.D.; Birdsey, R.; Finley, A.O.; Swantaran, A.; Dubayah, R.; Wayson, C.; Riemann, R. Integrating forest inventory and analysis data into a LIDAR-based carbon monitoring system. *Carbon Balance Manag.* **2014**, *9*, 1–11. [CrossRef]

219. Johnson, K.D.; Birdsey, R.; Cole, J.; Swatantran, A.; O'Neil-Dunne, J.; Dubayah, R.; Lister, A. Integrating LIDAR and forest inventories to fill the trees outside forests data gap. *Environ. Monit. Assess.* **2015**, *187*, 623. [CrossRef]

220. Sheridan, R.D.; Popescu, S.C.; Gatziolis, D.; Morgan, C.L.S.; Ku, N.-W. Modeling forest aboveground biomass and volume using airborne LiDAR metrics and Forest Inventory and Analysis data in the Pacific Northwest. *Remote Sens.* **2015**, *7*, 229–255. [CrossRef]

221. Hughes, R.F.; Asner, G.P.; Baldwin, J.A.; Mascaro, J.; Bufil, L.K.K.; Knapp, D.E. Estimating aboveground carbon density across forest landscapes of Hawaii: Combining FIA plot-derived estimates and airborne LiDAR. *Forest Ecol. Manag.* **2018**, *424*, 323–337. [CrossRef]

222. Joyce, M.J.; Erb, J.D.; Sampson, B.A.; Moen, R.A. Detection of coarse woody debris using airborne light detection and ranging (LiDAR). *Forest Ecol. Manag.* **2019**, *433*, 678–689. [CrossRef]

223. Lefsky, M.A.; Cohen, W.B.; Hudak, A.T.; Acker, S.A.; Ohmann, J.L. Integration of lidar, Landsat ETM+ and forest inventory data for regional forest mapping. *Int. Arch. Photogramm. Remote Sens.* **1999**, *32*, 119–126.

224. Chopping, M.; Schaaf, C.B.; Zhao, F.; Wang, Z.; Nolin, A.W.; Moisen, G.G.; Martonchik, J.V.; Bull, M. Forest structure and aboveground biomass in the southwestern United States from MODIS and MISR. *Remote Sens. Environ.* **2011**, *115*, 2943–2953. [CrossRef]

225. Deo, R.K.; Russell, M.B.; Domke, G.M.; Woodall, C.W.; Falkowski, M.J.; Cohen, W.B. Using Landsat time-series and LiDAR to inform aboveground forest biomass baselines in northern Minnesota, USA. *Can. J. Remote Sens.* **2017**, *43*, 28–47. [CrossRef]

226. Deo, R.K.; Russell, M.B.; Domke, G.M.; Andersen, H.-E.; Cohen, W.B.; Woodall, C.W. Evaluating site-specific and generic spatial models of aboveground forest biomass based on Landsat time-series and LiDAR strip samples in the eastern USA. *Remote Sens.* **2017**, *9*, 598. [CrossRef]

227. Deo, R.K.; Domke, G.M.; Russell, M.B.; Woodall, C.W.; Andersen, H.-E. Evaluating the influence of spatial resolution of Landsat predictors on the accuracy of biomass models for large-area estimation across the eastern USA. *Environ. Res. Lett.* **2018**, *13*, 055004. [CrossRef]

228. Ma, W.; Domke, G.M.; D'Amato, A.W.; Woodall, C.W.; Walters, B.F.; Deo, R.K. Using matrix models to estimate aboveground forest biomass dynamics in the eastern USA through various combinations of LiDAR, Landsat, and forest inventory data. *Environ. Res. Lett.* **2018**, *13*, 125004. [CrossRef]

229. Bell, D.M.; Gregory, M.J.; Kane, V.; Kane, J.; Kennedy, R.E.; Roberts, H.M.; Yang, Z. Multiscale divergence between Landsat- and lidar-based biomass mapping is related to regional variation in canopy cover and composition. *Carbon Balance Manag.* **2018**, *13*, 15. [CrossRef]

230. Gopalakrishnan, R.; Kauffman, J.S.; Fagan, M.E.; Coulston, J.W.; Thomas, V.A.; Wynne, R.H.; Fox, T.R.; Quirino, V.F. Creating landscape-scale site index maps for the southeastern U.S. is possible with airborne LiDAR and Landsat imagery. *Forests* **2019**, *10*, 234. [CrossRef]

231. Andersen, H.-E.; Barrett, T.; Winterberger, K.; Strunk, J.; Temesgen, H. Estimating forest biomass on the western lowlands of the Kenai Peninsula of Alaska using airborne lidar and field plot data in a model-assisted sampling design. In Proceedings of the IUFRO Division 4 Conference: "Extending Forest Inventory and Monitoring over Space and Time, Quebec, Canada, 19–22 May 2009; pp. 1–5.

232. Alonzo, M.; Morton, D.C.; Cook, B.D.; Andersen, H.-E.; Babcock, C.; Pattison, R. Patterns of canopy and surface layer consumption in a boreal forest fire from repeat airborne lidar. *Environ. Res. Lett.* **2017**, *12*, 065004. [CrossRef]

233. Strunk, J.L.; Reutebuch, S.E.; Andersen, H.-E.; Gould, P.J.; McGaughey, R.J. Model-assisted forest yield estimation with light detection and ranging. *West. J. Appl. For.* **2012**, *27*, 53–59. [CrossRef]

234. Strunk, J.; Temesgen, H.; Andersen, H.-E.; Flewelling, J.P.; Madsen, L. Effects of lidar pulse density and sample size on a model-assisted approach to estimate forest inventory variables. *Can. J. Remote Sens.* **2012**, *38*, 644–654. [CrossRef]

235. McRoberts, R.E.; Chen, Q.; Gormanson, D.D.; Walters, B.F. The shelf-life of airborne laser scanning data for enhancing forest inventory inferences. *Remote Sens. Environ.* **2018**, *206*, 254–259. [CrossRef]

236. Zwally, H.J.; Schutz, B.; Abdalati, W.; Abshire, J.; Bentley, C.; Brenner, A.; Bufton, J.; Dezio, J.; Hancock, D.; Harding, D.; et al. ICESat's laser measurements of polar ice, atmosphere, ocean, and land. *J. Geodyn.* **2002**, *34*, 405–445. [CrossRef]

237. Herzfeld, U.C.; McDonald, B.W.; Wallin, B.F.; Neumann, T.A.; Markus, T.; Brenner, A.; Field, C. Algorithm for detection of ground and canopy cover in micropulse photon-counting Lidar altimeter data in preparation for the ICESat-2 mission. *IEEE T. Geosci. Remote* **2014**, *52*, 2109–2125. [CrossRef]

238. Lefsky, M.A.; Harding, D.J.; Keller, M.; Cohen, W.B.; Carabajal, C.C.; Del Bom Espirito-Santo, F.; Hunter, M.O.; de Oliveira, R., Jr. Estimates of forest canopy height and aboveground biomass using ICESat. *Geophys. Res. Lett.* **2005**, *32*, L22S02. [CrossRef]

239. Pang, Y.; Lefsky, M.; Andersen, H.-E.; Miller, M.E.; Sherrill, K. Validation of the ICEsat vegetation product using crown-area-weighted mean height derived using crown delineation with discrete return lidar data. *Can. J. Remote Sens.* **2008**, *34*, S471–S484. [CrossRef]

240. Pflugmacher, D.; Cohen, W.; Kennedy, R.; Lefsky, M. Regional applicability of forest height and aboveground biomass models for the geoscience laser altimeter system. *For. Sci.* **2008**, *54*, 647–657.

241. Li, A.; Huang, C.; Sun, G.; Shi, H.; Toney, C.; Zhu, Z.; Rollins, M.G.; Goward, S.N.; Masek, J.G. Modeling the height of young forests regenerating from recent disturbances in Mississippi using Landsat and ICESat data. *Remote Sens. Environ.* **2011**, *115*, 1837–1849. [CrossRef]

242. Margolis, H.A.; Nelson, R.F.; Montesano, P.M.; Beaudoin, A.; Sun, G.; Andersen, H.-E.; Wulder, M.A. Combining satellite lidar, airborne lidar, and ground plots to estimate the amount and distribution of aboveground biomass in the boreal forest of North America. *Can. J. For. Res.* **2015**, *45*, 838–855. [CrossRef]

243. Nelson, R.; Margolis, H.; Montesano, P.; Sun, G.; Cook, B.; Corp, L.; Andersen, H.E.; deJong, B.; Pellat, F.P.; Fickel, T.; et al. Lidar-based estimates of aboveground biomass in the continental U.S. and Mexico using ground, airborne, and satellite observations. *Remote Sens. Environ.* **2017**, *188*, 127–140. [CrossRef]

244. Healey, S.P.; Patterson, P.L.; Saatchi, S.; Lefsky, M.A.; Lister, A.J.; Freeman, E.A. A sample design for globally consistent biomass estimation using lidar data from the Geoscience Laser Altimeter System (GLAS). *Carbon Balance Manag.* **2012**, *7*, 10. [CrossRef]

245. Dubayah, R.; Blair, J.B.; Goetz, S.; Fatoyinbo, L.; Hansen, M.; Healey, S.; Hofton, M.; Hurtt, G.; Kellner, J.; Luthcke, S.; et al. The global ecosystem dynamics investigation: High-resolution laser ranging of the Earth's forests and topography. *Sci. Remote Sens.* **2020**, *1*, 100002. [CrossRef]

246. Saarela, S.; Holm, S.; Healey, S.P.; Andersen, H.-E.; Petersson, H.; Prentius, W.; Patterson, P.L.; Næsset, E.; Gregoire, T.G.; Ståhl, G. Generalized hierarchical model-based estimation for aboveground biomass assessment using GEDI and Landsat data. *Remote Sens.* **2018**, *10*, 1832. [CrossRef]

247. Patterson, P.L.; Healey, S.P.; Ståhl, G.; Saarela, S.; Holm, S.; Andersen, H.-E.; Dubayah, R.O.; Duncanson, L.; Hancock, S.; Armston, J.; et al. Statistical properties of hybrid estimators proposed for GEDI—NASA's global ecosystem dynamics investigation. *Environ. Res. Lett.* **2019**, *14*, 065007. [CrossRef]

248. Gatziolis, D.; Lienard, J.F.; Vogs, A.; Strigul, N.S. 3D tree dimensionality assessment using photogrammetry and small unmanned aerial vehicles. *PLoS ONE* **2015**, *10*, e0137765. [CrossRef]

249. Farkhauser, K.E.; Strigul, N.S.; Gatziolis, D. Augmentation of traditional forest inventory and airborne laser scanning with unmanned aerial systems and photogrammetry for forest monitoring. *Remote Sens.* **2018**, *10*, 1562. [CrossRef]

250. Alonzo, M.; Andersen, H.-E.; Morton, D.; Cook, B. Quantifying boreal forest structure and composition using UAV structure from motion. *Forests* **2018**, *9*, 119. [CrossRef]

251. Strigul, N.; Gatziolis, D.; Liénard, J.; Vogs, A. Complementing forest inventory data with information from unmanned aerial vehicle imagery and photogrammetry. In *Proceedings of the 2015 Forest Inventory and Analysis (FIA) symposium, PNW-GTR-931, Portland, OR, USA, 8–10 December 2015*; Stanton, S., Christensen, G.A., Eds.; USDA Forest Service, Pacific Northwest Research Station: Portland, OR, USA, 2015; pp. 346–351.

252. Gatziolis, D.; Popescu, S.C.; Sheridan, R.; Ku, N.-W. Evaluation of terrestrial LiDAR technology forthe development of local tree volume equations. In *Proceedings of 10th International Conference on LiDAR Applications for Assessing Forest Ecosystems, Freiburg, Germany, 14–17 September 2010*; Koch, B., Kändler, G., Teguem, C., Eds.; University of Freiburg: Freiburg, Germany, 2010; pp. 197–205.

253. Klockow, P.A.; Putman, E.B.; Vogel, J.G.; Moore, G.W.; Edgar, C.B.; Popescu, S.C. Allometry and structural volume change of standing dead southern pine trees using non-destructive terrestrial LiDAR. *Remote Sens. Environ.* **2020**, *241*, 111729. [CrossRef]

254. McRoberts, R.E. Satellite image-based maps: Scientific inference or pretty pictures? *Remote Sens. Environ.* **2011**, *115*, 715–724. [CrossRef]

255. Reichenbach, H. *Experience and Prediction: An Analysis of the Foundations and the Structure of Knowledge*; University of Chicago Press: Chicago, IL, USA, 1938.

256. Bullock, E.L.; Woodcock, C.E.; Holden, C.E. Improved change monitoring using an ensemble of time series algorithms. *Remote Sens. Environ.* **2020**, *238*, 111165. [CrossRef]

257. Potapov, P.; Hansen, M.C.; Kommareddy, I.; Kommareddy, A.; Turubanova, S.; Pickens, A.; Adusei, B.; Tyukavina, A.; Ying, Q. Landsat analysis ready data for global land cover and land cover change mapping. *Remote Sens.* **2020**, *12*, 426. [CrossRef]

258. Moisen, G.G.; McConville, K.S.; Schroeder, T.A.; Healey, S.P.; Finco, M.V.; Frescino, T.S. Estimating land use and land cover change in north central Georgia: Can remote sensing observations augment traditional forest inventory data? *Forests* **2020**, *11*, 856. [CrossRef]

259. Patterson, P.L.; Coulston, J.W.; Roesch, F.A.; Westfall, J.A.; Hill, A.D. A primer of nonresponse in the U.S. Forest Inventory and Analysis program. *Env. Monit. Assess.* **2012**, *184*, 1423–1433. [CrossRef]

260. Schepaschenko, D.; See, L.; Lesiv, M.; Bastin, J.-F.; Mollicone, D.; Tsendbazar, N.-E.; Bastin, L.; McCallum, I.; Laso Bayas, J.C.; Baklanov, A.; et al. Recent advances in forest observation with visual interpretation of very high-resolution imagery. *Surv. Geophys.* **2019**, *40*, 839–862. [CrossRef]

261. USDA Forest Service. LCMS Data Explorer; USDA Forest Service. 2020. Available online: http://lcms.forestry.oregonstate.edu/tutorials/LCMS_Data_Explorer_Overview_20200317.pdf (accessed on 20 November 2020).

262. Tarrio, K.; Bullock, E.; Arevalo, P.; Turlej, K.; Zhang, K.; Holden, C.; Pasquarella, V.J.; Friedl, M.A.; Woodcock, C.E.; Olofsson, P.; et al. Global Land Cover mapping and Estimation (GLanCE): A multitemporal Landsat-based data record of 21st century global land cover, land use and land cover change. In *Poster Presentation, AGU Fall Meeting, 9–13 December 2019*; American Geophysical Union: San Francisco, CA, USA, 2019.

263. Gomes, V.C.F.; Queiroz, G.R.; Ferreira, K.R. An overview of platforms for big earth observation data management and analysis. *Remote Sens.* **2020**, *12*, 1253. [CrossRef]

264. Perry, C.; Oswalt, C.; Garner, J.; Bell, D.; Werstak, C.; Wilson, B. Public: Private partnerships improve the application of geographic information in decision-making. In Proceedings of the American Association of Geographers Annual Meeting, Washington, DC, USA, 3–7 April 2019.

265. Andersen, H.; Strunk, J.; Temesgen, H.; Atwood, D.; Winterberger, K. Using multilevel remote sensing and ground data to estimate forest biomass resources in remote regions: A case study in the boreal forests of interior Alaska. *Can. J. Remote Sens.* **2011**, *37*, 596–611. [CrossRef]

266. Sugarbaker, L.J.; Eldridge, D.F.; Jason, A.L.; Lukas, V.; Saghy, D.L.; Stoker, J.M.; Thunen, D.R. *Status of the 3D Elevation Program, 2015*; Open-File Report: Reston, VA, USA, 2017.

267. Irwin, J.R.; Sampath, A.; Kim, M.; Park, S.; Danielson, J.J. CountryMapper Sensor Validation Survey Data. U.S. Geological Survey, 2020. Available online: https://www.sciencebase.gov/catalog/item/5e5d679ae4b01d50924f2d23 (accessed on 13 December 2020).

268. Brown, M.E.; Delgado Arias, S.; Neumann, T.; Jasinski, M.F.; Posey, P.; Babonis, G.; Glenn, N.F.; Birkett, C.M.; Escobar, V.M.; Markus, T. Applications for ICESat-2 data: From NASA's early adopter program. *IEEE Geosci. Remote Sens. Mag.* **2016**, *4*, 24–37. [CrossRef]

269. Walker, W. An empirical InSAR-optical fusion approach to mapping vegetation canopy height. *Remote Sens. Environ.* **2007**, *109*, 482–499. [CrossRef]

270. Kellndorfer, J.M.; Walker, W.S.; LaPoint, E.; Kirsch, K.; Bishop, J.; Fiske, G. Statistical fusion of lidar, InSAR, and optical remote sensing data for forest stand height characterization: A regional-scale method based on LVIS, SRTM, Landsat ETM+, and ancillary data sets. *J. Geophys. Res. Biogeosci.* **2010**, *115*, G2. [CrossRef]

271. Yu, Y.; Saatchi, S.; Heath, L.S.; Lapoint, E.; Myneni, R.B.; Knyazikhin, Y. Regional distribution of forest height and biomass from multisensor data fusion. *J. Geophys. Res.* **2010**, *115*, 1–16. [CrossRef]

272. Kellndorfer, J.; Cartus, O.; Bishop, J.; Walker, W.; Holecz, F. Chapter 2: Large scale mapping of forests and land cover with synthetic aperture radar data. In *Land Applications of Radar Remote Sensing*; Holecz, F., Pasquali, P., Milisavljevic, N., Closson, D., Eds.; IntechOpen: London, UK, 2014; pp. 59–94.

273. Canty, M.J.; Nielsen, A.A.; Conradsen, K.; Skriver, H. Statistical analysis of changes in Sentinel-1 time series on the Google Earth Engine. *Remote Sens.* **2019**, *12*, 46. [CrossRef]

274. Flores, A.; Herndon, K.; Thapa, R.; Cherrington, E. *The SAR Handbook: Comprehensive Methodologies for Forest Monitoring and Biomass Estimation*; SERVIR Global Science Coordination Office: Huntsville, AL, USA, 2019.

275. McRoberts, R.; Wendt, D. Implementing a land cover stratification on-the-fly. In *Proceedings of the Third Annual Forest Inventory and Analysis Symposium, GTR NC-230, Traverse City, MI, USA, 17–19 October 2001*; McRoberts, R.E., Reams, G.A., Van Deusen, P.C., Moser, J.W., Eds.; U.S. Department of Agriculture, Forest Service, North Central Research Station: St. Paul, MN, USA, 2002; pp. 137–145.

276. Proctor, P.; Heath, L.; Deusen, P.; Gove, J.H.; Smith, J.E. COLE: A web-based tool for interfacing with Forest Inventory Data. In *Proceedings of the Fourth Annual Forest Inventory and Analysis Symposium, GTR-NC-252, New Orleans, LA, USA, 19–21 November 2002*; McRoberts, R.E., Reams, G.A., Van Deusen, C.P., McWilliams, W., Cieszewski, C., Eds.; U.S. Department of Agriculture, Forest Service, North Central Research Station: St. Paul, MN, USA, 2005; pp. 167–172.

277. Van Deusen, P.C.; Heath, L.S. Weighted analysis methods for mapped plot forest inventory data: Tables, regressions, maps and graphs. *For. Ecol. Manag.* **2010**, *260*, 1607–1612. [CrossRef]

278. McConville, K.; Tang, B.; Zhu, G.; Cheung, S.; Li, S. Mase: Model-Assisted Survey Estimation, 2018. Available online: https://cran.r-project.org/web/packages/mase/index.html (accessed on 13 December 2020).

279. R Core Team. *R: A Language and Environment for Statistical Computing, Version 3.5.1*; R Foundation for Statistical Computing: Vienna, Austria, 2018.

280. Molina, I.; Marhuendasae, Y. sae: An R package for small area estimation. *R J.* **2015**, *7*, 81–98. [CrossRef]

281. Breidenbach, J. JoSAE: Unit-Level and Area-Level Small Area Estimation, 2018. Available online: http://CRAN.R-project.org/package=JoSAE (accessed on 20 November 2020).

282. Riley, K.L.; Grenfell, I.C.; Finney, M.A. Mapping forest vegetation for the western United States using modified random forests imputation of FIA forest plots. *Ecosphere* **2016**, *7*, e01472. [CrossRef]

Publisher's Note: MDPI stays neutral with regard to jurisdictional claims in published maps and institutional affiliations.

 **forests**

Article

A Comparison of Forest Tree Crown Delineation from Unmanned Aerial Imagery Using Canopy Height Models vs. Spectral Lightness

Jianyu Gu *, Heather Grybas and Russell G. Congalton

Department of Natural Resources & the Environment, University of New Hampshire, 56 College Rd, Durham, NH 03824, USA; hmg10@wildcats.unh.edu (H.G.); russ.congalton@unh.edu (R.G.C.)
* Correspondence: gujianyu23@gmail.com

Received: 2 May 2020; Accepted: 23 May 2020; Published: 26 May 2020

Abstract: Improvements in computer vision combined with current structure-from-motion photogrammetric methods (SfM) have provided users with the ability to generate very high resolution structural (3D) and spectral data of the forest from imagery collected by unmanned aerial systems (UAS). The products derived by this process are capable of assessing and measuring forest structure at the individual tree level for a significantly lower cost compared to traditional sources such as LiDAR, satellite, or aerial imagery. Locating and delineating individual tree crowns is a common use of remotely sensed data and can be accomplished using either UAS-based structural or spectral data. However, no study has extensively compared these products for this purpose, nor have they been compared under varying spatial resolution, tree crown sizes, or general forest stand type. This research compared the accuracy of individual tree crown segmentation using two UAS-based products, canopy height models (CHM) and spectral lightness information obtained from natural color orthomosaics, using maker-controlled watershed segmentation. The results show that single tree crowns segmented using the spectral lightness were more accurate compared to a CHM approach. The optimal spatial resolution for using lightness information and CHM were found to be 30 and 75 cm, respectively. In addition, the size of tree crowns being segmented also had an impact on the optimal resolution. The density of the forest type, whether predominately deciduous or coniferous, was not found to have an impact on the accuracy of the segmentation.

Keywords: unmanned aerial systems; canopy height model; individual tree crown; segmentation

---

1. Introduction

Forests not only mitigate global climate change, sustain biodiversity, and prevent soil erosion; they also provide raw materials and resources such as timber, fresh food, and herbal medicines [1–3]. Maintaining the diversity of these products and services involves the development and implementation of forest management practices, which requires detailed forest inventory information at varying scales, such as stand-level basal area and diameter at breast height (DBH), and/or crown size and tree height at the single tree level [4–6].

The conventional way to gather this forest inventory information is to carry out periodic field surveys based on statistical sampling [7,8]. Nevertheless, the high cost in time and expense, as well as the difficulties in accessing specific sampling locations, make it an inefficient and often impractical approach [9,10]. Furthermore, data collected from in situ measurements, as shown in recent studies, is not as reliable due to many uncertainties such as sampling and observational errors [11–13]. Over the last few years, unmanned aerial systems (UAS), carrying a variety of sensors ranging from standard consumer-grade cameras to more expensive and complex multispectral or light detection and ranging

(LiDAR) sensors, have offered a potential solution to extend or replace field observations because of its ability to provide higher spatial resolution imagery and/or 3D data to quantify structural and compositional information at the single tree level [10]. This ability, combined with the tremendous progress in the techniques of digital image processing, has led to a sharp increase of these applications to precision forestry [14–16].

Individual tree locations and their crowns are the building blocks on which other parameters such as tree height, diameter at breast height (DBH), or biomass are estimated [17–19]. Treetops mark the tree locations and typical algorithms to detect them include local maximum filtering, image binarization, multiscale analysis, and template matching [12]. Methods to delineate tree crowns consist of three categories: valley following, watershed segmentation, and region growing [9,12]. The watershed algorithm, because of its intuitive and computationally efficient features, is one of the most commonly used segmentation algorithms for tree crown delineation. The algorithm metaphorically regards the whole grayscale image or model as a topographic surface where the watershed lines are the boundaries of trees [20,21]. However, due to its high sensitivity to noise and spectral variation, it is prone to oversegmentation, a situation where multiple segments fall within what should be a single tree crown [22]. Many improved watershed algorithms such as edge-embedded, marker-controlled, or multiscale approaches were developed to overcome this problem [23,24]. The marker-controlled watershed algorithm, which adds marker regions or points corresponding to one segmented object, was shown to be robust and flexible [20,25,26]. Many studies successfully applied a marker-controlled watershed to delineate tree crowns and achieved accuracies over 85% [22,27].

The data for detecting treetops or segmenting individual tree crowns could be derived either photogrammetrically or from LiDAR [12,28,29]. Digital photogrammetry is favored by many researchers to calculate forest inventory metrics because of its ability to provide orthometrically corrected imagery (orthoimagery) in addition to 3D point clouds for a much lower price compared to a LiDAR system. The point clouds derived from photogrammetry are extracted from stereo images based on structure-from-motion (SfM) and multiview stereopsis (MVS) techniques [30–32]. However, unlike the LiDAR-based point cloud, because of its inability to penetrate the foliage to achieve the ground information, it can only generate a digital surface model (DSM) for dense forests [33]. An external digital terrain model (DTM) is usually needed to create the canopy height model (CHM) representing the height of objects above the ground. Either the orthoimagery or CHM can then be used for tree segmentation [28,34]. Most research developed algorithms assuming that tree canopies possess mountainous shape, where treetops are the locally brightest in the image or the locally highest in the CHM data, while tree edges are darker or lower in elevation [12,35]. Very little research has examined and compared the accuracies of tree crowns segmented from UAS and photogrammetrically-based imagery and CHMs, especially within dense coniferous and deciduous forest stands.

Data that are photogrammetrically generated are of exceptionally high spatial resolution (e.g., pixel size of a few centimeters) but often provide too much detail. For example, tree branches and gaps between leaves increases the spectral or height variation within the tree crown, adding to the uncertainty for tree crown segmentation [22,36]. Upscaling, decreasing the spatial resolution of the original data, is one way to reduce this noise, but it can also weaken the distinction between tree crowns [12,22]. Additionally, canopies of different sizes may have varying degrees of sensitivity to the chosen spatial resolution. Intuitively, as the spatial resolution decreases, the segmentation accuracy of the larger crowns may increase because potential noise within the crown is reduced. In contrast, the accuracy of smaller crowns declines because they may disappear in coarser images [12]. Therefore, a tradeoff exists between the tree size and spatial resolution; thus, it would benefit users to find the best spatial resolution for specific tree crown sizes.

The objectives of this study are to (1) compare the accuracies of individual tree crowns delineated from UAS-based CHMs and natural color orthoimagery using maker-controlled watershed segmentation, (2) provide insight into how accuracies change with spatial resolution, crown size, and forest type (coniferous or deciduous), and (3) facilitate the consideration of choosing the right data and scale for individual tree delineation in the future.

2. Data and Methodology

2.1. Study Site and Data Collection

This research took place within the College Woods Natural Area (CWNA, Figure 1), 70°56′51.339″ W and 43°8′7.935′ N, in Durham, NH, USA. The CWNA is owned and managed by the University of New Hampshire. The annual average precipitation for the region is 119.38 cm with a yearly average temperature of 8.84 °C. Two soil types, Buxton and Hollis–Charlton, dominate this area. White pine (*Pinus strobus*), eastern hemlock (*Tsuga canadensis*), American beech (*Fagus grandifolia*), and several species of oak (*Quercus* spp.) are the primary tree species/genera. Two study sites, each covering a 400 × 400 m area, were chosen within the CWNA. Both study sites are a mixed forest type; however, the coniferous tree species are most prevalent in study site #1, while the deciduous tree species are dominant in study site #2.

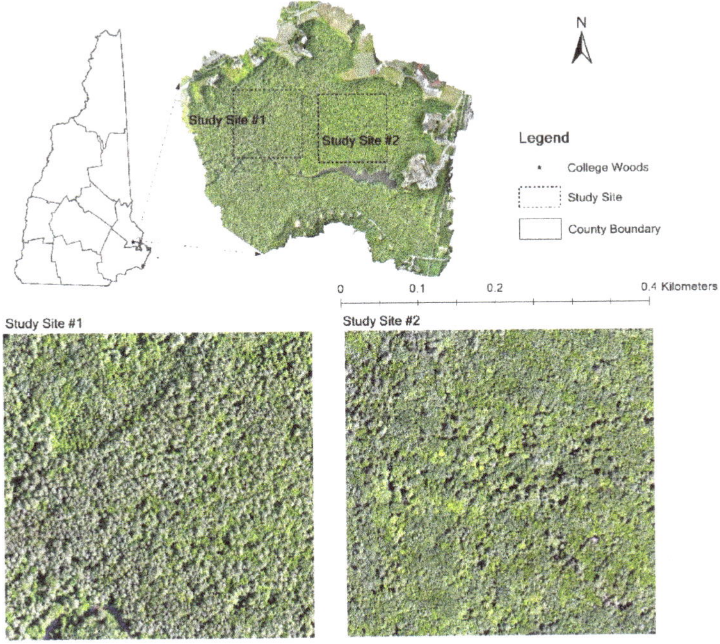

Figure 1. Study sites at College Woods, New Hampshire, USA

The raw UAS images were collected on 11 July 2018, with a fixed-winged SenseFly eBee Plus carrying a SODA (sensor optimized for digital acquisition) camera that captures natural color imagery. The flight was 120 m above the ground with a forward and side overlap of 85%. A total of 1961 photos, covering all of the CWNA, were collected.

2.2. Data Preprocessing

The first step of preprocessing was to create an orthomosaic and DSM from the UAS imagery. All the raw images were processed with Sensefly's Flight Data Manager built into the eMotion software [37]. First, the geotags for all the images collected during the mission were extracted from the mission flight log and Post-Processed Kinematic (PPK) processed using a nearby Continuously

Operating Reference Station (CORS) (site ID: NHUN). The images were then geotagged with the PPK corrected positions. Due to the density of the canopy cover, ground control points could not be collected across the CWNA. The images were further processed by Agisoft Metashape Pro (v.1.6.2) [38] to create a natural orthomosaic and DSM. The Agisoft workflow comprises five basic steps: align photos, build dense cloud, build mesh, build digital elevation model, and build orthomosaic [39]. We followed the suggestions provided by [40] to set the parameters for data processing. The spatial resolution of the orthomosaic and DSM was 2.31 and 12.10 cm, respectively.

The second step was to create a series of data sets with different spatial resolutions from the orthomosaic and CHM to test the effects of spatial resolution. The orthomosaic was converted from an RGB model into an HSL model, where the lightness band (L) represents pixel brightness. The lightness band is widely utilized for object segmentation [41,42]. The CHM model was created by subtracting a DTM from the UAS-based DSM. The DTM was made from LiDAR data collected for coastal New Hampshire in the winter and spring of 2011, and downloaded from the GRANIT LiDAR distribution site [43]. The 2-meter gridded raster DTM files, generated from ground-classified LiDAR returns, were provided as part of the project deliverables. Based on the size and land-use history of two study sites, the age of the DTM relative to the UAS missions would introduce little, if any, error. The DTM was reprojected to match the projection, coordinate system, and horizontal and vertical datum of the DSM. A series of datasets with different spatial resolutions were created by resampling the lightness band and CHM using cubic convolution in ArcGIS Pro 2.4.2 [44]. For the lightness band, the resampling started at 2 cm and was incrementally increased by 2 cm until a resolution of 100 cm was reached, resulting in 50 lightness datasets. The same process was performed on the CHM; however, the initial resolution was 12 cm, resulting in 44 CHMs.

2.3. Reference Tree Crown Data Collection

The reference data (i.e., individual reference tree crowns) were randomly collected from each study site and then manually interpreted by combining the natural color orthomosaic and CHM data. The workflow follows. First, 800 random sampling points were generated over each study site. Then, a trained undergraduate student manually digitized a tree if a point fell within the tree's crown. If more than two points were situated within the same tree crown, only one tree crown was counted in. Any point that hit the background (not a tree) was removed. However, the edges of the canopy are highly curved, making digitizing work extremely arduous. In order to reduce the workload without losing the accuracy of reference data, an extremely oversegmented result was created by applying the multiresolution segmentation algorithm in eCognition 9.5.1 [45]. The scale, compactness, and shape parameters for the algorithm were set to 40, 0.5, and 0.5, respectively. Finally, the interpreter digitized the tree crown by merging the crowns' oversegmented polygons into a single crown polygon. A few polygons may have still needed a splitting operation before merging, but this workflow improved the processing of delineating the reference data.

Another experienced researcher further examined the interpreted result, and all controversial objects were removed after discussion. The final reference tree crown polygons for a study site were divided into three groups, large, medium, and small trees, based on the crown area using natural division. For study site #1, the criteria of separation were: large ($\geq$42.06 m^2), medium (18.42–42.06 m^2), and small (<18.42 m^2). For study site #2, the criteria of separation were: large ($\geq$51.20 m^2), medium (21.74–51.20 m^2), and small (<21.74 m^2). The sample size in each group was uneven. To make all groups comparable, we randomly resampled all other groups without replacement using a sample size based on the group with the least number of samples across both study sites. That group was the large trees in Site #1 with only 174 reference trees.

2.4. Treetops Detection

This research applied a local maximum filter to detect the treetops which is highly dependent on crown size [46]. The window size was determined by calculating the average size of the tree crowns in the reference samples. The window size was set to 4.58 and 4.51 m for study site #1 and #2, respectively.

2.5. Marker-Controlled Watershed Segmentation

The watershed algorithm is a classical algorithm for segmentation, which was developed from mathematical morphology [47]. The marker-controlled watershed algorithm requires two inputs: (1) a gray scale image to represent the "topography" or highs and lows of the area, and (2) the point locations (i.e., markers) that define either local minimums or maximums within the gray scale image [48]. When the markers represent local minimums, the algorithm delineates a polygon around each marker containing higher gradient (i.e., spectrally brighter or topographically higher) pixels than that of the marker. In this research, local maximums representing treetops were used as the markers, inverting the processes so the delineated areas represent a decreasing gradient of values around the treetop. The area delineated around the marker in this case was assumed to represent that tree's crown. The markers act as seed locations for the algorithm and, unlike traditional watershed segmentation, restricts the creation of basins to just those markers. This creates a one-to-one relationship between markers and segments or trees and crowns. Details of the marker-controlled watershed algorithm can be found in [20,49,50].

A Sobel filter, a widely used algorithm, was applied to each dataset to calculate gradients [51]. The marker-controlled watershed segmentation was applied to all the lightness bands and CHMs using scikit-image, an open source image processing library for the Python programming language [52]. It is worth noting that during the workflow, smoothing filters such as the Gaussian filter were not applied across the data to reduce noise because these filters are regarded as having a similar effect as reducing the spatial resolution. The combined operations would weaken the purpose of this research to explore the best scale for segmentation.

2.6. Accuracy Assessment

The accuracy assessment for segmentation is different from the one for traditional thematic classification [53]. The purpose of individual tree crown delineation is to represent each crown with a single polygon [12]. Therefore, before calculating the accuracy measures for each reference polygon, the best-matching segment from each segmentation result must be chosen to build a one-to-one relationship. The overlap index (OI) proposed by [54] was utilized in this research to find the single best candidate for each reference polygon.

$$OI = \frac{area\ (r_i \cap s_j)}{area\ (r_i)} \times \frac{area\ (r_i \cap s_j)}{area\ (s_j)} \tag{1}$$

In Equation (1), r_i represents i^{th} reference polygon and s_j represents the jth candidate segmented polygon that intersects with r_i. The symbol $\cap$ represents the intersection of r_i and s_j. OI ranges from 0 to 1, where a higher value indicates a better match.

This research employed oversegmentation accuracy (*Oa*), undersegmentation accuracy (*Ua*), and quality rate (*QR*) to quantitatively validate the segmentation results [55,56].

$$Oa = \frac{1}{n} \sum_{i=1}^{n} \left(\frac{area\ (r_i \cap s_i)}{area\ (r_i)} \right) \tag{2}$$

$$Ua = \frac{1}{n} \sum_{i=1}^{n} \left(\frac{area\ (r_i \cap s_i)}{area\ (s_i)} \right) \tag{3}$$

In Equations (2) and (3), the s_i indicates the best corresponding candidate. The sampling size is represented by n. A higher *Oa* or *Ua* means greater accuracy. The *QR* proposed by [57] defines the accuracy between a reference polygon and its candidate by combining the overlapped and union region. It also considers the geometrical similarity. If a segmented object entirely coincides with its reference object, the *QR* reaches the minimum [56].

$$QR = \frac{1}{n} \sum_{i=1}^{n} \left(1 - \frac{area\ (r_i \cap s_i)}{area\ (r_i \cup s_i)} \right)$$

(4)

In Equation (4), the $\cup$ denotes a union. Higher *QR* indicates a less accurate segmentation.

3. Results

Figure 2 presents the Oa, Ua, and QR of all segmentations using the lightness band as the data source for study site #1. The accuracies are displayed for four groups: large, medium, small, and all crowns, as follows:

(1) For large crowns, the Ua is higher than the Oa. Overall, the gap between Ua and Oa is narrower when the spatial resolution reaches between 16 and 72 cm. The Ua shows a downward trend while the Oa demonstrates an upward trend before the spatial resolution approaches 74 cm. Both the lines of Ua and Oa become stable when the spatial resolution is between 16 and 48 cm. The highest Ua is approximately 0.81 when the spatial resolution is 2 cm. The Oa reaches a maximum value of 0.62 when spatial resolution is 68 cm. The QR shows a general downward trend before the spatial resolution of 74 cm. The QR lies under 0.6 for spatial resolutions between 26 and 48 cm, and'between 58 and 72 cm. As indicated by the minimum of QR, the best segmentation is achieved when the spatial resolution is 46 cm.

(2) The lines of Oa and Ua for medium crowns intertwine, and the gap between them becomes narrower in contrast to the one in the large group. It results in a stable QR around 0.60. The lowest QR appears when spatial resolution is 54 cm.

(3) The three accuracy measures for the small group are quite different. The line for Oa is much higher than the one for Ua. The gap between them becomes narrower after the spatial resolution reaches 74 cm. All Ua values are under 0.50. Most QR values are higher than 0.70, which is higher than the ones in either the large or medium groups.

(4) Both the shape and values of Oa and Ua for all crowns parallel the medium crowns. The QR value varies between 0.60 and 0.70. The relative lower QR values appear when the spatial resolution lies between 30 and 62 cm.

Figure 3 presents the accuracies after segmenting the CHM and exhibits a clear difference from Figure 2. First, the lines of Oa, Ua, and QR are highly stable for all crown sizes. Most values of Ua and Oa are lower than the ones in Figure 2, resulting in higher QR values. Second, within each group, the line of Oa is higher than the Ua except in the case of large crowns. The Ua reduces, and the Oa increases as the crown size grows. According to QR, the best spatial resolution for segmentation is 86, 78, 74, and 76 cm for the large, medium, small, and all groups, respectively.

Figure 4 demonstrates the results from study site #2 using the lightness band as the segmenting data. A similar trend is shown as in Figure 2. The minor difference is that the values of Oa are lower with higher Ua, resulting in a broader gap in the large group. The best spatial resolution for segmentation is 68, 58, 2, and 30 cm for the large, medium, small, and all groups, respectively.

The results in Figure 5 resemble those in Figure 3, and the differences between Figures 4 and 5 are similar to those between Figures 2 and 3. The best spatial resolution for segmentation is 100, 74, 74, and 74 cm for the large, medium, small, and all groups, respectively.

Table 1 further shows the average accuracy measures for all spatial resolutions. Regardless of the study sites and groups, the mean QR value is lower with a higher Ua using the lightness band as a data source compared to using CHM, although mean Oa in the small group is slightly lower. When using the lightness band, the Oa is higher with lower Ua and QR if comparing study site #1 to study #2. However, there is little difference between them in each group using CHM.

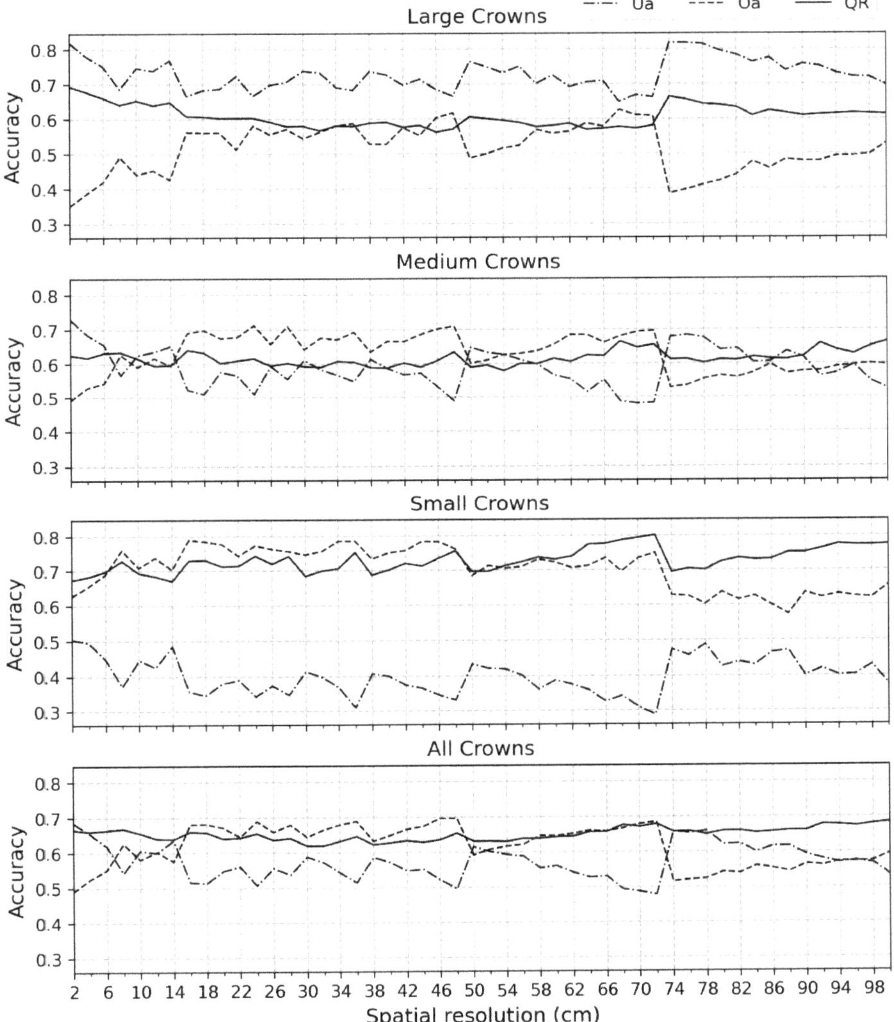

Figure 2. Impact of spatial resolution and crown size on Oa, Ua, and QR using the lightness band in study site #1.

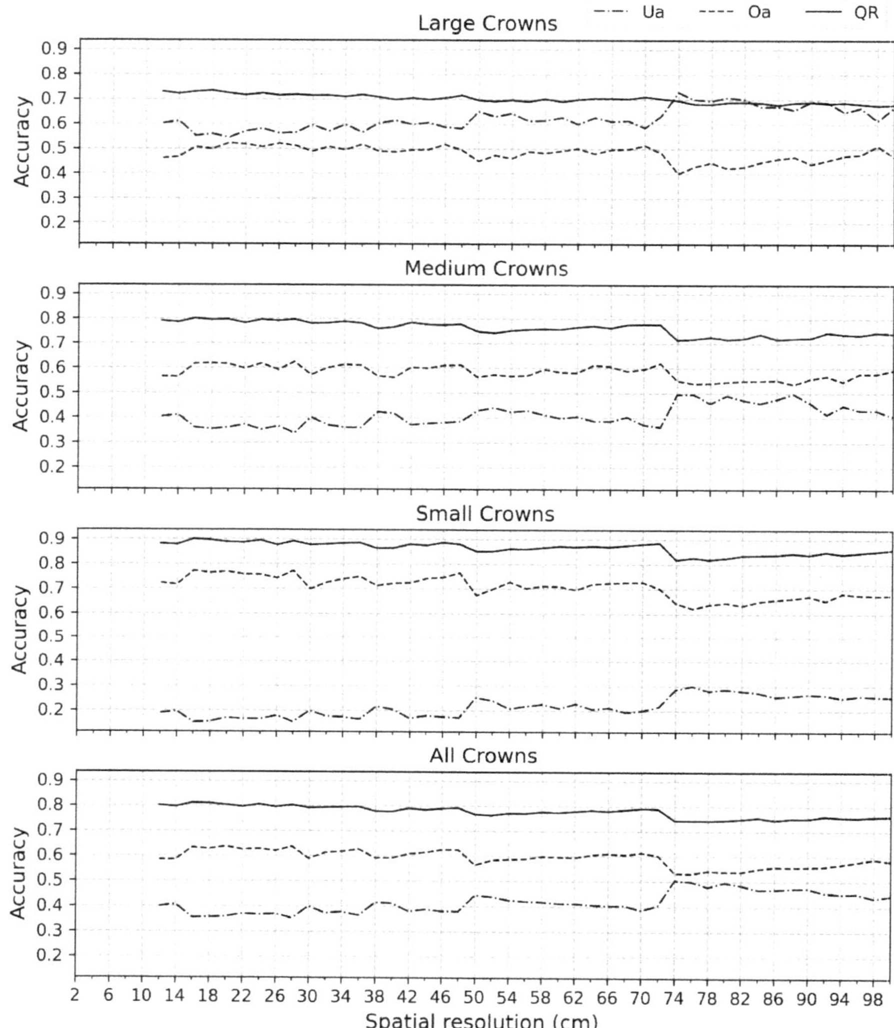

Figure 3. Impact of spatial resolution and crown size on Oa, Ua, and QR using the canopy height model (CHM) in study site #1.

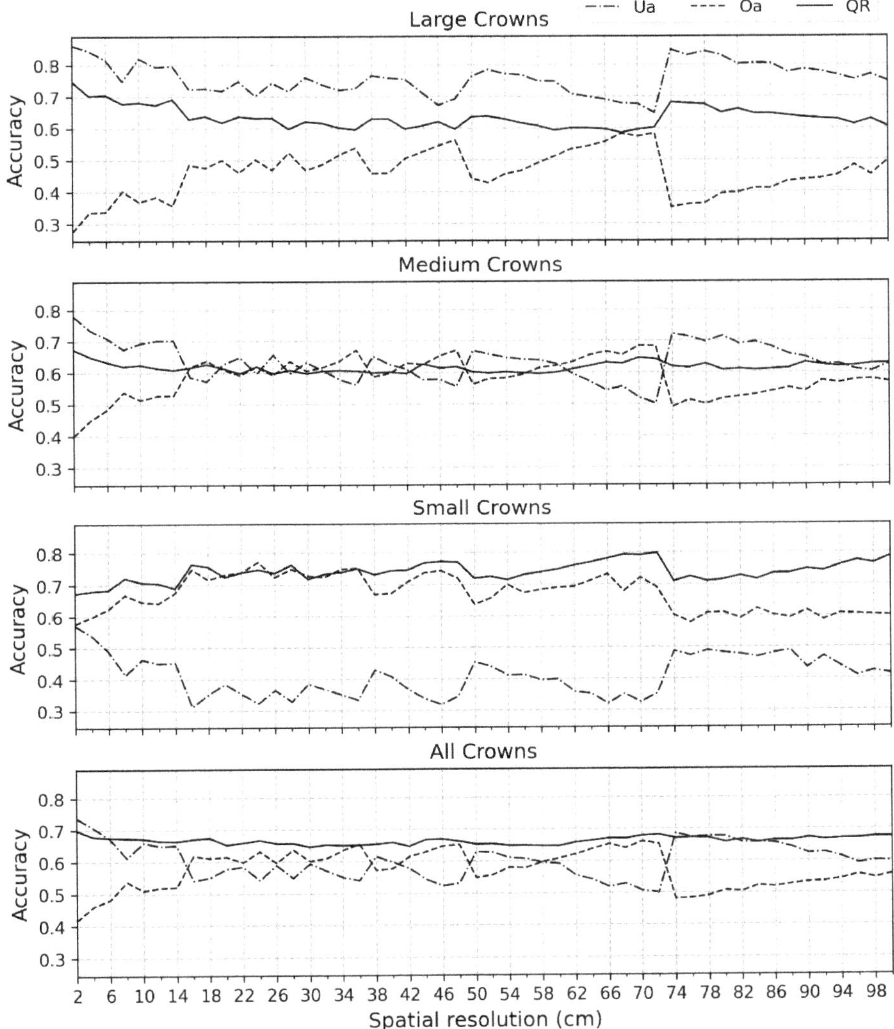

Figure 4. Impact of spatial resolution and crown size on Oa, Ua, and QR using the lightness band in study site #2.

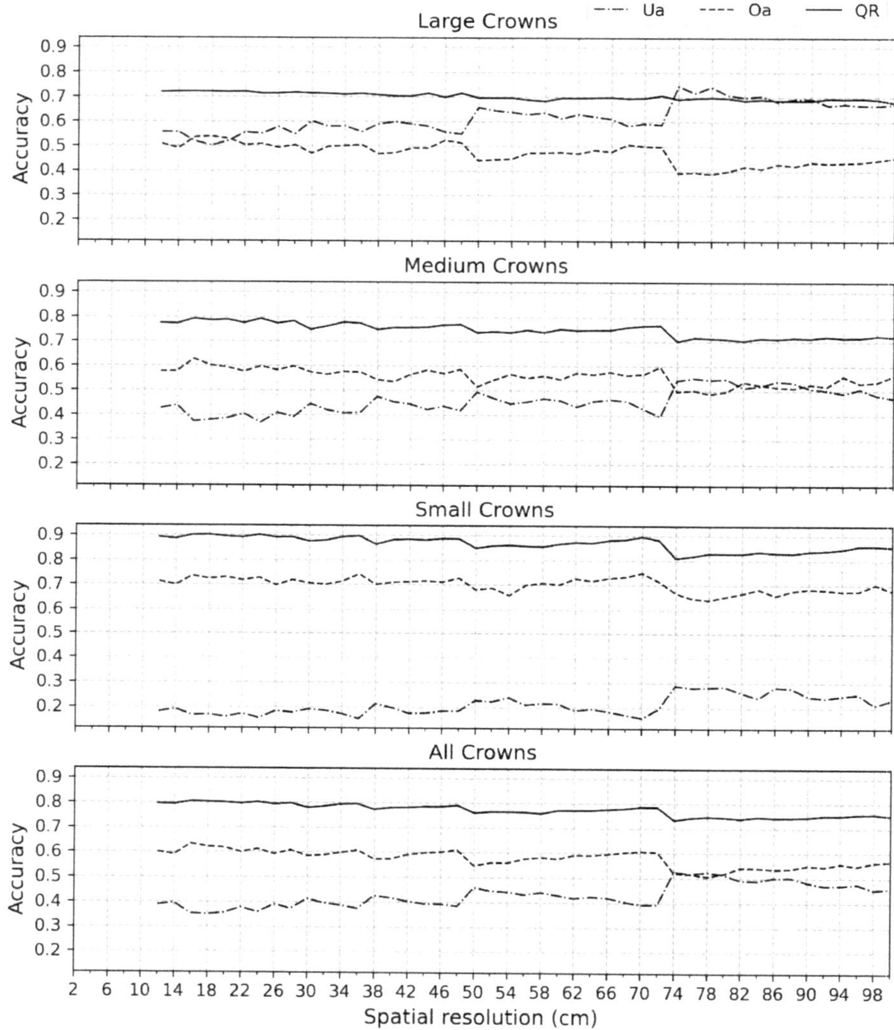

Figure 5. Impact of spatial resolution and crown size on Oa, Ua, and QR using the CHM in study site #2.

Table 1. Average accuracy measures of all spatial resolutions using lightness or CHM from each study site.

Group	Study Site #1						Study Site #2					
	Lightness			CHM			Lightness			CHM		
	Oa	Ua	QR	Oa	Ua	QR	Oa	Ua	QR	Oa	Ua	QR
Large	0.51	0.73	0.61	0.48	0.62	0.70	0.46	0.76	0.63	0.47	0.62	0.70
Medium	0.63	0.59	0.62	0.58	0.41	0.76	0.58	0.63	0.62	0.56	0.46	0.75
Small	0.70	0.40	0.73	0.71	0.22	0.86	0.67	0.41	0.74	0.70	0.21	0.87
All	0.62	0.57	0.65	0.59	0.42	0.78	0.57	0.60	0.66	0.57	0.43	0.77

4. Discussion

This research examined and compared the accuracy of segmenting individual tree crowns from CHMs and spectral lightness bands using maker-controlled watershed segmentation. Additionally, the effects of spatial resolution, crown size, and forest type on delineation accuracy were also investigated. The Ua, Oa, and QR are widely accepted for validating segmentation and were reported as accuracy measures in this study.

This research demonstrates that single tree crowns segmented from the lightness band are more accurate than those segmented from the CHM if both data were derived from digital photogrammetry (Figures 2–5 and Table 1)). The underlying reason is the low quality of the CHM impacted by, for example, data source or geoprocessing [30]. First, the point cloud produced through the SfM algorithm has limited ability to detect the small gaps and peaks in the crown, which gives rise to an underestimation of the upper layers of the canopy but an overestimation of the lower layers [58,59]. Second, the edges of crowns are usually darker, lower, and often obscured by surrounding trees and are, therefore, less visible in the imagery compared to higher parts of the crowns, including the treetops [32]. The SfM–MVS process relies on the computer being able to "see" features in the imagery in order to generate a 3D position (point) [60]. Fewer points would be created at the edges, which results in a relatively smoothed and underestimated DSM based on interpolation. Both these factors would lead to an undersegmented result, which is confirmed by the fact that differences between lightness band–Ua and CHM–Ua are higher than the ones between lightness band–Oa and CHM–Oa in each study site. Third, due to the characteristics of dense forest in both study sites, digital photogrammetry can only produce the point cloud from the canopy surface visible to the camera [61]. An external DTM is needed to calculate the CHM; however, the inconsistency of the spatial resolution becomes a factor [39]. Previous research focused on comparing the CHM from the LiDAR to the one derived from digital photogrammetry based on SfM [58]. This research complements the comparison between lightness and CHM, with both from digital photogrammetry. We prove that watershed segmentation using a CHM is less accurate for a dense forest than using the natural color images and suggest that a systematic error budget analysis of CHMs derived from photogrammetry based on SfM is necessary.

Results show that spatial resolution alters the accuracy of segmentation. It is worth noting that the spectral properties of the downscaled images will not be the same as an image captured with a native spatial resolution matching that of the downscaled image (i.e., an image downscaled from 2 to 30 cm is not the same as an image captured at 30 cm to begin with). However, small UAS in the United States are not legally allowed to fly higher than 122 m (legally 400 feet) above the ground and thus the maximum pixel size that can be achieved is restricted by flying height and the sensor's properties. The best spatial resolution both for study site #1 and #2 using lightness is located at 30 cm. Comparable accuracies lie between 30 and 62 cm, and between 26 and 42 cm, respectively. The best spatial resolution for segmentation using CHM for study site #1 and #2 is 76 and 74 cm, respectively; however, the variation of accuracies due to spatial resolution is more stable. These results provide a basis for how to adopt the best spatial resolution or kernel size for smoothing filters in the future. This research also confirms that as the spatial resolution decreases, the segmentation of the large, medium, and small crowns reaches its best accuracy at various scales, which provides the implications for segmenting trees of particular interest (e.g., large trees). However, this conclusion is limited by defining the size of trees, which is usually determined by the diameter at breast height (DBH). Although the allometric function to estimate DBH from canopy width was explored in Japan by Iizuka, Yonehara, Itoh, and Kosugi [39], such a local equation does not exist for the study area.

Based on the average QR, the segmentation accuracy does not differ much between study site #1 and #2, although study site #1 has higher Oa but lower Ua. Unlike the coniferous trees, which typically follow a distinct mountainous shape, the canopies of deciduous trees are usually much flatter [12]. Multiple treetops are prone to be detected within the deciduous crown, resulting in an oversegmentation problem, which is very obvious in the large and medium crowns using lightness as the data source (Figures 4 and 5, and Table 1). The minor difference in QR between study site #1 and #2 implies that the density of the forest exerts more influence on the segmentation accuracy rather than the forest type. Besides, the reconstruction of the point cloud is limited by the smoothing in the dense matching process, creating abrupt and discontinuous vertical changes in the CHM, especially for the coniferous trees in the mixed forest [30,58]. Although research on detection and segmentation of deciduous trees has increased [62–64], segmenting deciduous trees in high density stands based on UAS imagery is still under development.

This research also implies that the size of the sampling reference objects impacts segmentation accuracy assessment (Figures 2 and 4). Previous research favored the stratified random sampling for traditional thematic classification [53,65], but the sampling design for segmentation accuracy remains unresolved [66] and which attributes (e.g., size or shape) are best for stratified sampling needs further study.

5. Conclusions

This research compared the use of a CHM with the lightness band for the delineation of individual tree crowns based on the maker-controlled watershed algorithm. It also examined how segmentation accuracy varies due to spatial resolution, crown size, and forest type. The study highlights the following conclusions. The single tree crowns segmented from the lightness band based on the marker-control watershed algorithm are more accurate than those using the CHM if both data are derived from digital photogrammetry. The best spatial resolution using lightness is 30 cm, with comparable scales between 26 and 62 cm. The best spatial resolution for segmentation using a CHM is around 75 cm. The large trees are prone to be oversegmented, while the small trees are prone to be undersegmented. The best spatial resolution for segmenting trees of different size varies. Mixed forest type dominated by either deciduous or coniferous does not show much difference in accuracy. Finally, this research suggests that the size of the reference polygons impacts segmentation accuracy assessment, which deserves more investigation in the future.

Author Contributions: J.G., H.G., and R.G.C. conceived and designed the experiments. J.G. performed the experiments and analyzed the data with guidance from R.G.C.; J.G. wrote the paper. H.G. and R.G.C. edited and finalized the paper and manuscript. All authors have read and agreed to the published version of the manuscript.

Funding: Partial funding was provided by the New Hampshire Agricultural Experiment Station. This is Scientific Contribution Number: #2856. This work was supported by the USDA National Institute of Food and Agriculture McIntire Stennis Project #NH00095-M (Accession #1015520).

Acknowledgments: The authors would like to acknowledge Vincent Pagano, Hannah Stewart, and Benjamin Fraser for their assistance with the reference data collection.

Conflicts of Interest: The authors declare no conflict of interest.

References

1. Devi, R.M.; Patasaraiya, M.K.; Sinha, B.; Saran, S.; Dimri, A.P.; Jaiswal, R. Understanding the linkages between climate change and forest. *Curr. Sci.* **2018**, *114*, 987–996. [CrossRef]
2. Mitchard, E.T.A. The tropical forest carbon cycle and climate change. *Nature* **2018**, *559*, 527–534. [CrossRef] [PubMed]
3. Llopart, M.; Reboita, M.S.; Coppola, E.; Giorgi, F.; da Rocha, R.P.; de Souza, D.O. Land Use Change over the Amazon Forest and Its Impact on the Local Climate. *Water* **2018**, *10*, 149. [CrossRef]
4. Ling, P.Y.; Baiocchi, G.; Huang, C.Q. Estimating annual influx of carbon to harvested wood products linked to forest management activities using remote sensing. *Clim. Chang.* **2016**, *134*, 45–58. [CrossRef]
5. Rodriguez-Gonzalez, P.M.; Albuquerque, A.; Martinez-Almarza, M.; Diaz-Delgado, R. Long-term monitoring for conservation management: Lessons from a case study integrating remote sensing and field approaches in floodplain forests. *J. Environ. Manag.* **2017**, *202*, 392–402. [CrossRef]
6. Vauhkonen, J.; Imponen, J. Unsupervised classification of airborne laser scanning data to locate potential wildlife habitats for forest management planning. *Forestry* **2016**, *89*, 350–363. [CrossRef]

7.   Tomppo, E.; Malimbwi, R.; Katila, M.; Mäkisara, K.; Henttonen, H.M.; Chamuya, N.; Zahabu, E.; Otieno, J. A sampling design for a large area forest inventory: Case Tanzania. *Can. J. For. Res.* **2014**, *44*, 931–948. [CrossRef]

8.   Bergseng, E.; Orka, H.O.; Naesset, E.; Gobakken, T. Assessing forest inventory information obtained from different inventory approaches and remote sensing data sources. *Ann. For. Sci.* **2015**, *72*, 33–45. [CrossRef]

9.   Wagner, F.H.; Ferreira, M.P.; Sanchez, A.; Hirye, M.C.M.; Zortea, M.; Gloor, E.; Phillips, O.L.; de Souza Filho, C.R.; Shimabukuro, Y.E.; Aragao, L.E.O.C. Individual tree crown delineation in a highly diverse tropical forest using very high resolution satellite images. *ISPRS J. Photogramm. Remote Sens.* **2018**, *145*, 362–377. [CrossRef]

10. Chen, G.; Weng, Q.; Hay, G.J.; He, Y. Geographic object-based image analysis (GEOBIA): Emerging trends and future opportunities. *GISci. Remote Sens.* **2018**, *55*, 159–182. [CrossRef]

11. Wang, Y.; Lehtomäki, M.; Liang, X.; Pyörälä, J.; Kukko, A.; Jaakkola, A.; Liu, J.; Feng, Z.; Chen, R.; Hyyppä, J. Is field-measured tree height as reliable as believed—A comparison study of tree height estimates from field measurement, airborne laser scanning and terrestrial laser scanning in a boreal forest. *ISPRS J. Photogramm. Remote Sens.* **2019**, *147*, 132–145. [CrossRef]

12. Ke, Y.; Quackenbush, L.J. A review of methods for automatic individual tree-crown detection and delineation from passive remote sensing. *Int. J. Remote Sens.* **2011**, *32*, 4725–4747. [CrossRef]

13. Sačkov, I.; Santopuoli, G.; Bucha, T.; Lasserre, B.; Marchetti, M. Forest Inventory Attribute Prediction Using Lightweight Aerial Scanner Data in a Selected Type of Multilayered Deciduous Forest. *Forests* **2016**, *7*, 307. [CrossRef]

14. Ke, Y.; Quackenbush, L.J.; Im, J. Synergistic use of QuickBird multispectral imagery and LIDAR data for object-based forest species classification. *Remote Sens. Environ.* **2010**, *114*, 1141–1154. [CrossRef]

15. Pu, R.; Landry, S. A comparative analysis of high spatial resolution IKONOS and WorldView-2 imagery for mapping urban tree species. *Remote Sens. Environ.* **2012**, *124*, 516–533. [CrossRef]

16. Hansen, M.C.; Potapov, P.V.; Goetz, S.J.; Turubanova, S.; Tyukavina, A.; Krylov, A.; Kommareddy, A.; Egorov, A. Mapping tree height distributions in Sub-Saharan Africa using Landsat 7 and 8 data. *Remote Sens. Environ.* **2016**, *185*, 221–232. [CrossRef]

17. Ploton, P.; Barbier, N.; Couteron, P.; Antin, C.M.; Ayyappan, N.; Balachandran, N.; Barathan, N.; Bastin, J.F.; Chuyong, G.; Dauby, G.; et al. Toward a general tropical forest biomass prediction model from very high resolution optical satellite images. *Remote Sens. Environ.* **2017**, *200*, 140–153. [CrossRef]

18. Yilmaz, V.; Yilmaz, C.S.; Tasci, L.; Gungor, O. Determination of Tree Crown Diameters with Segmentation of a UAS-Based Canopy Height Model. *IPSI BGD Trans. Internet Res.* **2017**, *13*, 63–67.

19. Liu, G.J.; Wang, J.L.; Dong, P.L.; Chen, Y.; Liu, Z.Y. Estimating Individual Tree Height and Diameter at Breast Height (DBH) from Terrestrial Laser Scanning (TLS) Data at Plot Level. *Forests* **2018**, *9*, 398. [CrossRef]

20. Gaetano, R.; Masi, G.; Poggi, G.; Verdoliva, L.; Scarpa, G. Marker-Controlled Watershed-Based Segmentation of Multiresolution Remote Sensing Images. *IEEE Trans. Geosci. Remote Sens.* **2015**, *53*, 2987–3004. [CrossRef]

21. Chen, S.; Luo, J.C.; Shen, Z.F.; Hu, X.D.; Gao, L.J. Segmentation of Multi-spectral Satellite Images Based on Watershed Algorithm. In Proceedings of the 2008 International Symposium on Knowledge Acquisition and Modeling, Wuhan, China, 21–22 December 2008; pp. 684–688. [CrossRef]

22. Huang, H.Y.; Li, X.; Chen, C.C. Individual Tree Crown Detection and Delineation from Very-High-Resolution UAV Images Based on Bias Field and Marker-Controlled Watershed Segmentation Algorithms. *IEEE J. Sel. Top. Appl. Earth Obs. Remote Sens.* **2018**, *11*, 2253–2262. [CrossRef]

23. Cai, Y.Q.; Tong, X.H.; Shu, R.; IEEE. Multi-scale Segmentation of Remote Sensing Image Based on Watershed Transformation. In Proceedings of the 2009 Joint Urban Remote Sensing Event, Shanghai, China, 20–22 May 2009; p. 425. [CrossRef]

24. Li, D.R.; Zhang, G.F.; Wu, Z.C.; Yi, L.N. An Edge Embedded Marker-Based Watershed Algorithm for High Spatial Resolution Remote Sensing Image Segmentation. *IEEE Trans. Image Process.* **2010**, *19*, 2781–2787. [CrossRef] [PubMed]

25. Mylonas, S.; Stavrakoudis, D.; Theocharis, J.; Mastorocostas, P. A Region-Based GeneSIS Segmentation Algorithm for the Classification of Remotely Sensed Images. *Remote Sens.* **2015**, *7*, 2474–2508. [CrossRef]

26. Wang, M.; Li, R. Segmentation of High Spatial Resolution Remote Sensing Imagery Based on Hard-Boundary Constraint and Two-Stage Merging. *IEEE Trans. Geosci. Remote Sens.* **2014**, *52*, 5712–5725. [CrossRef]

27. Fang, F.; Im, J.; Lee, J.; Kim, K. An improved tree crown delineation method based on live crown ratios from airborne LiDAR data. *Gisci. Remote Sens.* **2016**, *53*, 402–419. [CrossRef]
28. Mohan, M.; Silva, C.A.; Klauberg, C.; Jat, P.; Catts, G.; Cardil, A.; Hudak, A.T.; Dia, M. Individual Tree Detection from Unmanned Aerial Vehicle (UAV) Derived Canopy Height Model in an Open Canopy Mixed Conifer Forest. *Forests* **2017**, *8*, 340. [CrossRef]
29. Wallace, L.; Lucieer, A.; Watson, C.S. Evaluating Tree Detection and Segmentation Routines on Very High Resolution UAV LiDAR Data. *IEEE Trans. Geosci. Remote Sens.* **2014**, *52*, 7619–7628. [CrossRef]
30. Lisein, J.; Pierrot-Deseilligny, M.; Bonnet, S.; Lejeune, P. A Photogrammetric Workflow for the Creation of a Forest Canopy Height Model from Small Unmanned Aerial System Imagery. *Forests* **2013**, *4*, 922–944. [CrossRef]
31. Planck, N.R.V.; Finley, A.O.; Kershaw, J.A.; Weiskittel, A.R.; Kress, M.C. Hierarchical Bayesian models for small area estimation of forest variables using LiDAR. *Remote Sens. Environ.* **2018**, *204*, 287–295. [CrossRef]
32. Wallace, L.; Lucieer, A.; Malenovsky, Z.; Turner, D.; Vopenka, P. Assessment of Forest Structure Using Two UAV Techniques: A Comparison of Airborne Laser Scanning and Structure from Motion (SfM) Point Clouds. *Forests* **2016**, *7*, 62. [CrossRef]
33. Zhang, J.; Lin, X. Advances in fusion of optical imagery and LiDAR point cloud applied to photogrammetry and remote sensing. *Int. J. Image Data Fusion* **2017**, *8*, 1–31. [CrossRef]
34. Wang, M.; Dong, Z.P.; Cheng, Y.F.; Li, D.R. Optimal Segmentation of High-Resolution Remote Sensing Image by Combining Superpixels With the Minimum Spanning Tree. *IEEE Trans. Geosci. Remote Sens.* **2018**, *56*, 228–238. [CrossRef]
35. Li, Z.; Hayward, R.; Zhang, J.; Liu, Y. Individual Tree Crown Delineation Techniques for Vegetation Management in Power Line Corridor. In Proceedings of the 2008 Digital Image Computing: Techniques and Applications, Canberra, ACT, Australia, 1–3 December 2008; pp. 148–154.
36. Milas, A.S.; Arend, K.; Mayer, C.; Simonson, M.A.; Mackey, S. Different colours of shadows: Classification of UAV images. *Int. J. Remote Sens.* **2017**, *38*, 3084–3100. [CrossRef]
37. SenseFly User-Manuals. Available online: https://www.sensefly.com/my-sensefly/user-manuals/ (accessed on 29 April 2020).
38. Agisoft Metashape User Manual Professional Edition 1.6. Available online: https://www.agisoft.com/pdf/metashape-pro_1_6_en.pdf (accessed on 29 April 2020).
39. Iizuka, K.; Yonehara, T.; Itoh, M.; Kosugi, Y. Estimating Tree Height and Diameter at Breast Height (DBH) from Digital Surface Models and Orthophotos Obtained with an Unmanned Aerial System for a Japanese Cypress (Chamaecyparis obtusa) Forest. *Remote Sens.* **2018**, *10*, 13. [CrossRef]
40. Fraser, B.T.; Congalton, R.G. Issues in Unmanned Aerial Systems (UAS) Data Collection of Complex Forest Environments. *Remote Sens.* **2018**, *10*, 908. [CrossRef]
41. Poblete-Echeverría, C.; Olmedo, G.; Ingram, B.; Bardeen, M. Detection and Segmentation of Vine Canopy in Ultra-High Spatial Resolution RGB Imagery Obtained from Unmanned Aerial Vehicle (UAV): A Case Study in a Commercial Vineyard. *Remote Sens.* **2017**, *9*, 268. [CrossRef]
42. Chen, Y.; Hou, C.; Tang, Y.; Zhuang, J.; Lin, J.; He, Y.; Guo, Q.; Zhong, Z.; Lei, H.; Luo, S. Citrus Tree Segmentation from UAV Images Based on Monocular Machine Vision in a Natural Orchard Environment. *Sensors* **2019**, *19*, 5558. [CrossRef] [PubMed]
43. GRANIT LiDAR Distribution Site. Available online: http://lidar.unh.edu/map/ (accessed on 29 April 2020).
44. ESRI ArcGIS Pro 2.4.2. Available online: https://www.esri.com/en-us/arcgis/products/arcgis-pro/resources (accessed on 21 May 2020).
45. Belgiu, M.; Dragut, L. Comparing supervised and unsupervised multiresolution segmentation approaches for extracting buildings from very high resolution imagery. *ISPRS J. Photogramm. Remote Sens.* **2014**, *96*, 67–75. [CrossRef] [PubMed]
46. Wulder, M.; Niemann, K.O.; Goodenough, D.G. Local Maximum Filtering for the Extraction of Tree Locations and Basal Area from High Spatial Resolution Imagery. *Remote Sens. Environ.* **2000**, *73*, 103–114. [CrossRef]
47. Zhang, Y.; Feng, X.; Le, X. Segmentation on Multispectral Remote Sensing Image Using Watershed Transformation. In Proceedings of the 2008 Congress on Image and Signal Processing, Hainan, China, 27–30 May 2008; pp. 773–777.
48. Kornilov, S.A.; Safonov, V.I. An Overview of Watershed Algorithm Implementations in Open Source Libraries. *J. Imaging* **2018**, *4*, 123. [CrossRef]

49. Polewski, P.; Yao, W.; Cao, L.; Gao, S. Marker-free coregistration of UAV and backpack LiDAR point clouds in forested areas. *ISPRS J. Photogramm. Remote Sens.* **2019**, *147*, 307–318. [CrossRef]

50. Xiao, P.F.; Feng, X.Z.; Zhao, S.H.; Me, S.J.; Wang, P.F.; Badawi, R. Applying texture marker-controlled watershed transform to the segmentation of IKONOS image. In *Geoinformatics 2007: Remotely Sensed Data and Information, Pts 1 and 2*; Ju, W., Zhao, S., Eds.; SPIE: Bellingham, WA, USA, 2007; Volume 6752.

51. Furnari, A.; Farinella, G.M.; Bruna, A.R.; Battiato, S. Distortion adaptive Sobel filters for the gradient estimation of wide angle images. *J. Vis. Commun. Image Represent.* **2017**, *46*, 165–175. [CrossRef]

52. van der Walt, S.; Schönberger, J.L.; Nunez-Iglesias, J.; Boulogne, F.; Warner, J.D.; Yager, N.; Gouillart, E.; Yu, T. scikit-image: Image processing in Python. *PeerJ* **2014**, *2*, e453. [CrossRef] [PubMed]

53. Congalton, R.G.; Green, K. *Assessing the Accuracy of Remotely Sensed Data: Principles and Practices*, 3rd ed.; CRC Press: Boca Raton, FL, USA, 2019.

54. Yang, J.; He, Y.; Caspersen, J. Region merging using local spectral angle thresholds: A more accurate method for hybrid segmentation of remote sensing images. *Remote Sens. Environ.* **2017**, *190*, 137–148. [CrossRef]

55. Möller, M.; Lymburner, L.; Volk, M. The comparison index: A tool for assessing the accuracy of image segmentation. *Int. J. Appl. Earth Obs. Geoinf.* **2007**, *9*, 311–321. [CrossRef]

56. Chen, Y.Y.; Ming, D.P.; Zhao, L.; Lv, B.R.; Zhou, K.Q.; Qing, Y.Z. Review on High Spatial Resolution Remote Sensing Image Segmentation Evaluation. *Photogramm. Eng. Remote Sens.* **2018**, *84*, 629–646. [CrossRef]

57. Weidner, U. Contribution to the assessment of segmentation quality for remote sensing applications. *Int. Arch. Photogramm. Remote Sens.* **2008**, *37*, 479–484.

58. Jayathunga, S.; Owari, T.; Tsuyuki, S. Evaluating the Performance of Photogrammetric Products Using Fixed-Wing UAV Imagery over a Mixed Conifer–Broadleaf Forest: Comparison with Airborne Laser Scanning. *Remote Sens.* **2018**, *10*, 187. [CrossRef]

59. Vastaranta, M.; Wulder, M.; White, J.; Pekkarinen, A.; Tuominen, S.; Ginzler, C.; Kankare, V.; Holopainen, M.; Hyyppä, J.; Hyyppä, H. Airborne laser scanning and digital stereo imagery measures of forest structure: Comparative results and implications to forest mapping and inventory update. *Can. J. Remote Sens.* **2013**, *39*, 382–395. [CrossRef]

60. Iglhaut, J.; Cabo, C.; Puliti, S.; Piermattei, L.; O'Connor, J.; Rosette, J. Structure from Motion Photogrammetry in Forestry: A Review. *Curr. For. Rep.* **2019**, *5*, 155–168. [CrossRef]

61. Matese, A.; Di Gennaro, S.F.; Berton, A. Assessment of a canopy height model (CHM) in a vineyard using UAV-based multispectral imaging. *Int. J. Remote Sens.* **2017**, *38*, 2150–2160. [CrossRef]

62. Nuijten, J.G.R.; Coops, C.N.; Goodbody, R.H.T.; Pelletier, G. Examining the Multi-Seasonal Consistency of Individual Tree Segmentation on Deciduous Stands Using Digital Aerial Photogrammetry (DAP) and Unmanned Aerial Systems (UAS). *Remote Sens.* **2019**, *11*, 739. [CrossRef]

63. Ene, L.; Næsset, E.; Gobakken, T. Single tree detection in heterogeneous boreal forests using airborne laser scanning and area-based stem number estimates. *Int. J. Remote Sens.* **2012**, *33*, 5171–5193. [CrossRef]

64. Hamraz, H.; Contreras, M.A.; Zhang, J. A robust approach for tree segmentation in deciduous forests using small-footprint airborne LiDAR data. *Int. J. Appl. Earth Obs. Geoinf.* **2016**, *52*, 532–541. [CrossRef]

65. Stehman, S.V. Sampling designs for accuracy assessment of land cover. *Int. J. Remote Sens.* **2009**, *30*, 5243–5272. [CrossRef]

66. Ye, S.; Pontius, R.G.; Rakshit, R. A review of accuracy assessment for object-based image analysis: From per-pixel to per-polygon approaches. *ISPRS J. Photogramm. Remote Sens.* **2018**, *141*, 137–147. [CrossRef]

MDPI

St. Alban-Anlage 66

4052 Basel

Switzerland

Tel. +41 61 683 77 34

Fax +41 61 302 89 18

www.mdpi.com

Forests Editorial Office

E-mail: forests@mdpi.com

www.mdpi.com/journal/forests